T0305157

Payments for Environmental Services, Forest Conservation and Climate Change

Payments for Environmental Services, Forest Conservation and Climate Change

Livelihoods in the REDD?

Edited by

Luca Tacconi

The Australian National University, Canberra

Sango Mahanty

The Australian National University, Canberra

Helen Suich

The Australian National University, Canberra

Edward Elgar

Cheltenham, UK • Northampton, MA, USA

Published by
Edward Elgar Publishing Limited
The Lypiatts
15 Lansdown Road
Cheltenham
Glos GL50 2JA
UK

Edward Elgar Publishing, Inc.
William Pratt House
9 Dewey Court
Northampton
Massachusetts 01060
USA

A catalogue record for this book
is available from the British Library

Library of Congress Number: 2010927664

ISBN 978 1 84980 299 4

Typeset by Servis Filmsetting Ltd, Stockport, Cheshire
Printed and bound by MPG Books Group, UK

Contents

Contributors

Dallay Annawi uses her skills in the social sciences, applied sociology and anthropology in conducting research with Environmental Science for Social Change (ESSC), a Jesuit research organization in the Philippines that works toward greater human security and environmental sustainability. Over the past 10 years, she has been involved in ESSC's various projects: assistance on ancestral domain management planning with Kankanaey indigenous communities in Mountain Province; promoting regional knowledge-sharing on community-based natural resource management; tracking mining discussions; and increasing disaster resilience.

Hilza Domingos Silva dos Santos Arcos graduated from the Federal University of Acre (UFAC) as a geographer and holds a specialization in Water Resource Management from the University of Taubaté, São Paulo. Born in Acre, she currently works for PESACRE (Group of Research and Extension in Agroforestry Systems of Acre), an NGO that conducts research and extension for integrated conservation and development in the Amazon. Using participatory methods she provides technical assistance to smallholder family producers to motivate the adoption of sustainable production practices, strengthen social organization and manage conflict. She is a member of the Council for Sustainable Rural Development for the Alto Acre Region and Capixaba (CTAC) and co-author of *Planejando futuros sustentáveis para os pequenos produtores: Programa Proambiente Polo Alto Acre* (2008).

Wendy-Lin Bartels works with the South East Climate Consortium and the Florida Climate Institute at the University of Florida. Her current research involves engaging stakeholder groups and decisionmakers in Florida's coastal and agricultural sectors as they prepare for and adapt to a changing and variable climate. Wendy-Lin also collaborates with partners in the Brazilian Amazon, where she previously studied a multistakeholder land-use planning process that encourages small-scale producers to implement sustainable practices that provide environmental services. Her broad career goal is to better understand the factors that facilitate dialogue among diverse stakeholders as they negotiate the impacts of climate change. Wendy-Lin has a PhD in Interdisciplinary

Ecology with a specialization in Tropical Conservation and Development, a Master's in Science Communication and a Bachelor of Science in Botany and Molecular Genetics.

Eduardo Amaral Borges is an agronomist who arrived in the north-western Brazilian Amazon in 1992. He is currently the coordinator of PESACRE (Group of Research and Extension in Agroforestry Systems of Acre), an NGO that conducts research and extension for integrated conservation and development in the Amazon. Since 1999, he has worked with small-holder family producers through participatory methods and technical assistance to motivate the adoption of sustainable production practices, strengthen social organization and manage conflict. He is a member of the National Council for Food Security and Nutrition (CONSEA) and author of *Plano de Uso dos Recursos Naturais do PDS São Salvador* (2003) and *Como produzir farinha d'água com qualidade* (1998).

Esteve Corbera is a Senior Research Associate at the School of International Development, University of East Anglia, and the Tyndall Centre for Climate Change Research in the United Kingdom. His research focuses on the governance of clean development and land-use, including analyses of Clean Development Mechanism projects, voluntary offset markets, PES initiatives and other climate policy instruments, such as Reducing Emissions from Deforestation and Forest Degradation. He is a member of the International Society for Ecological Economics and a fellow of the Foundation for a New Ethical Business.

Adair Pereira Duarte is a technical agent in environmental management and graduated from the University of the North (Centro Universitário do Norte/UNINORTE) with a specialization in Auditing and Environmental Accounting. He is a native of São Roque, Paraná, and currently works with PESACRE (Group of Research and Extension in Agroforestry Systems of Acre), an NGO that conducts research and extension for integrated conservation and development in the Amazon. Using participatory methods he provides technical assistance to smallholder family producers to motivate the adoption of sustainable production practices, strengthen social organization, and manage conflict. He is a member of the Network of Agroecology and co-author of *Planejando futuros sustentáveis para os pequenos produtores: Programa Proambiente Polo Alto Acre* (2008).

Laura A. German is a Senior Scientist at the Centre for International Forestry Research (CIFOR) in Bogor, Indonesia, and Leader of CIFOR's research on 'Managing Impacts of Globalised Trade and Investment on

Forests and Forest Communities.' Past work includes action research on integrated natural resource management and landscape governance in eastern Africa; human ecological research in the Brazilian Amazon; and agricultural development in Honduras. She holds a PhD in Cultural and Ecological Anthropology from the University of Georgia and a BSc in Engineering from Cornell University.

Pablo Gutman is Environmental Economics Director at the World Wildlife Fund in Washington DC. For the past 30 years Mr Gutman has worked, researched and lectured on issues of environment and development in over 20 countries in Latin America, Africa and Asia, and has published widely in journals, books and collective works. Mr Gutman holds an MSc on Development Economics from Buenos Aires University and an MSc on Environmental Economics from London University.

Jonathan Haskett is a climate change scientist whose work focuses on the nexus of land use, climate change mitigation, adaptation and poverty reduction in the developing world. His work spans the creation of land use carbon projects, the development of landscape carbon measurement methodologies and climate change policy at the domestic and international levels. Dr Haskett holds a PhD in Soil Science from the University of Minnesota and has done post-doctoral research on the effect of climate change on agriculture with USDA-ARS. He helped in the development of the long-term acquisition plan for the Landsat-7 satellite while at the University of Maryland. As a Peace Corps volunteer in Ecuador (2002–2005) Dr Haskett worked with indigenous communities on soil conservation, agroforestry and mapping of the landscape carbon resource. Since 2007 he has been with the World Agroforestry Centre, a position that included close collaboration with the World Wildlife Fund.

Rohit Jindal is a Doctoral Candidate, with specialization in Environmental Economics, at Michigan State University. He also holds an MSc in Resource Management from the University of Edinburgh. He has conducted field research on PES and natural resource projects in India, Mozambique, Kenya and Tanzania. He is presently working on the use of field auctions in removing information asymmetry between buyers and sellers of ecosystem services, as part of a research fellowship from the World Agroforestry Centre. He has won several awards, including the 2nd prize in the World Resources Institute's EarthTrends global environmental essay competition.

Charlotte Kalanzi is an environmentalist. She is currently assisting the Uganda Wildlife Authority to gather community knowledge and experience of ongoing gas and oil exploration activities close to Wildlife Management Areas in Uganda. Charlotte worked with the Katoomba Group in a survey of the potential of the private sector in East Africa to engage in markets and payments for ecosystem services. Charlotte also took part in identifying the livelihood impacts of incentive payments for Reducing Emissions from Deforestation and Degradation (REDD) in Bushenyi District, Uganda. She holds a degree in Environmental Management from Makerere University.

Beria Leimona is currently a PhD Candidate at Wageningen Agricultural University in the Netherlands. She also works for the World Agroforestry Centre on the second phase of the Rewards for Use of and Shared Investment in Pro-poor Environmental Services project. She specializes in environmental services and community-based natural resource management issues.

Sango Mahanty specializes in forest governance and rural development, with experience in the South Pacific, South Asia and mainland Southeast Asia. Her areas of expertise include collaborative resource management, participatory development, social assessment, participatory monitoring and evaluation, and the livelihood implications of resource management and development programs. For the past three years, she has researched the implications of payment for environmental service schemes and REDD for rural livelihoods and poverty. She also coordinates the Master of Applied Anthropology and Participatory Development programme at the Australian National University, Canberra.

Richard Mwesigwa holds an MSc in Environment and Natural Resources Management from Makerere University's Institute of Environment and Natural Resources and a BA in Sociology and Geography also from Makerere University, Kampala. He works for Nature Harness Initiatives (NAHI) as a Programs Officer, and doubles as a researcher in both socio-economic and conservation studies. He has been involved in developing and promoting market-based approaches to conservation of productive landscapes in Uganda. He has some years' experience in research with particular interest in PES, community livelihoods, policy and institutional frameworks, community access and use and management of natural resources, both on private land and in protected areas.

Stefano Pagiola is a Senior Environmental Economist in the Latin America and Caribbean Sustainable Development Department of the World Bank. He leads the World Bank's work on Payments for Environmental Services

(PES) and has helped develop projects implementing the approach in numerous countries, as well as publishing extensively in this area and on the valuation of environmental benefits. He received a BA from Princeton University and an MA and PhD from Stanford University. He taught environmental economics at Stanford University before joining the World Bank in 1994.

Rachman Pasha is a research officer for the World Agroforestry Centre, South East Asia. He is currently pursuing his Master's Degree at Bogor University of Agriculture.

N.P. Rahadian is an Executive Director of Rekonvasi Bhumi, an NGO pioneering the development of payments for ecosystem services in Serang, Banten Province, Indonesia.

Ana R. Rios works on agricultural development, particularly in topics related to natural resources, productivity, markets and poverty. Ana obtained a BSc in Agricultural Economics and Horticulture from Escuela Agricola Panamericana Zamorano (2000), and then received an MSc (2003) and PhD (2008) in Agricultural Economics from Purdue University. She has been a consultant to the World Bank (2005) and a post-doctoral research associate at the Centre for Global Trade Analysis (2008). She is currently a research fellow at the Inter-American Development Bank.

Alice Ruhweza is a private consultant, who currently coordinates the East and Southern Africa Katoomba Group, a regional working group of individuals interested in advancing environmental markets. Before joining private consultancy, she coordinated Uganda's National Environment Management Authority Lead Agency Component, which ensured environmental concerns were integrated into policies and plans of line ministries and private businesses. Alice has also worked with Sprint Corporation, and the African Environmental Research and Consulting Group. She holds a Bachelor's Degree in Economics from Makerere University and a Master's Degree in Applied Economics from the University of Wisconsin, USA.

Marianne Schmink is Professor of Latin American Studies and Anthropology at the University of Florida, where she is Director of the Tropical Conservation and Development (TCD) program. She has co-authored or edited several books and over 50 articles, book chapters, and reports. Since 1986, she has directed a major collaborative program in Acre with the Federal University of ACRE (UFAC) and non-government organizations including PESACRE and CTA. She has worked on issues related to gender, development and community-based conservation for over 20 years.

Rowena Soriaga has been conducting research and facilitating knowledge sharing on different concerns relating to development management in rural Asia over the last 15 years. She currently manages the secretariat operations of the Asia Forest Network, a coalition of organizations and individuals dedicated to promoting the role of local communities in sustainable management and the restoration of Asia's forests. With an educational background in development management and business economics, she also contributes to technical working group discussions on equitable and sustainable development at sub-national, national and international levels.

Helen Suich is a development and resource economist. She works on issues relating to poverty alleviation, sustainable economic development and natural resource management. She is currently a Doctoral Candidate at the Crawford School of Economics and Government, the Australian National University, Canberra.

Luca Tacconi is Director of the Environmental Management and Development Program at the Crawford School of Economics and Government, the Australian National University, Canberra. His research focuses on the economic, political and social factors that drive environmental change – resulting in loss of biodiversity and climate change – and their implications for rural livelihoods and poverty.

Acronyms

AFOLU	agriculture, forest and other land use
BR&D	BioClimate Research and Development
CBFM	community-based forest management
CCBS	Climate, Community and Biodiversity Standard
CDI	National Commission for the Development of Indigenous Peoples
CDM	Clean Development Mechanism
CENRO	Community Environment and Natural Resources Office
CER	Certified Emission Reductions
CNA	National Water Commission
CONAFOR	National Forestry Commission
CONGEN	federal level Management Council (*Proambiente*)
CTOs	Certifiable Tradable Offsets
DENR	Department of Environment and Natural Resources
DILG	Department of Interior and Local Government
DPWH	Department of Public Works and Highways
ECCM	Edinburgh Centre for Carbon Management
EcoTrust	Environment Conservation Trust of Uganda
ES	environmental service(s)
ESI	environmental services index
EU	European Union
FGDs	focus group discussions
FKDC	*Forum Komunikasi DAS Cidanau*
FONAFIFO	National Forest Finance Fund (*Fondo Nacional de Financiamiento Forestal*)
FUNDECOR	Foundation for the Development of the Central Volcanic Mountain Range
GEF	Global Environment Facility
GNP	Gorongosa National Park
GTZ	German Technical Cooperation (*Deutsche Gesellschaft für Technische Zusammenarbeit*)
ha	hectare
INE	National Institute of Ecology
LULUCF	Land Use, Land Use Change and Forestry
MEs	microenterprises

xiii

MINAE	Ministry of the Environment and Energy
NFB	'No-Fire Bonus' scheme
NGO	non-government organization
OLS	ordinary least squares
PES	Payments for environmental/ecosystem services
PESACRE	Proambiente implementing agency in the state of Acre
PIN	Project Idea Note (for the Clean Development Mechanism)
PSA	*Pago por Servicios Ambientales* (Spanish for PES)
PSA-CABSA	Payments for Carbon, Biodiversity and Agroforestry Services
PSA-H	Payments for Forest's Hydrological Environmental Services Programme (*Program de Pagos por Servicios Ambientales Hidrológicos de los Bosques*)
PT KTI	PT Krakatau Tirta Industry
REDD	reduced emissions for deforestation and forest degradation
TA	technical assistance
TAC	Technical Advisory Council
tCO$_2$e	tons of carbon dioxide equivalent
UNDP	United Nations Development Programme
UNEP	United Nations Environment Programme
UNFCCC	United Nations Framework Convention on Climate Change
VCS	Voluntary Carbon Standard

Acknowledgements

The research presented in this book was supported by funding from the Australian Development Agency for International Development through the Australian Development Research Awards, project EFCC083 'Assessing the livelihood impacts of incentive payments for avoided deforestation'. Other funding sources that contributed to the research presented in the individual chapters are acknowledged there.

We would like to thank the following colleagues for reviewing chapters of the book: Simon Anstey, James Blignaut, Esteve Corbera, Florence Daviet, Rohit Jindal, Marshall Murphree, Charles Palmer and Sven Wunder. Many thanks also to Chris Ulyatt for his editing work.

1. Forests, payments for environmental services and livelihoods

Luca Tacconi, Sango Mahanty and Helen Suich

INTRODUCTION

Payment for Environmental Services (PES) schemes, where the providers of environmental services receive payments for the adoption of land uses and practices that support those services, are relatively recent in the developing world. There is strong interest in PES schemes because of their potential to mobilize new resources for conservation and achieve development outcomes. This interest has increased with recent discussions under the United Nations Framework Convention on Climate Change (UNFCCC) on a mechanism for Reduced Emissions from Deforestation and forest Degradation (REDD).[1] REDD would require the provision of financial incentives to developing countries to conserve their forests, and could possibly include payments to people with rights over the forests in question. These developments have heightened interest in learning from past and present PES schemes, matched by concerns about their impacts – and those of REDD – on the rights and livelihoods of local resource users and managers.

This book therefore addresses the following questions:

- What have been the impacts of PES schemes on livelihoods?
- What are the implications for the design of incentive mechanisms for REDD at the local level?

To address these questions, case studies of PES schemes were selected to represent implementation at a variety of scales, with different tenure structures, across Africa, Asia and Latin America. This book presents these case studies, and concludes with a comparison of the main livelihood impacts documented in these case studies and the implications for the design of incentive mechanisms for REDD at the local level.

This chapter introduces the need for research on the livelihood impacts of PES, and discusses the methodological framework guiding the research. It discusses the concept of REDD and the link with PES, reviews literature on PES and livelihoods, then presents the framework and questions guiding the individual case studies. It then provides an overview of the case studies presented in the following chapters.

REDUCED EMISSIONS FROM DEFORESTATION AND DEGRADATION AND LIVELIHOODS

Deforestation and forest degradation[2] contribute about 12 per cent of global anthropogenic CO_2 emissions, with peatland conversion and degradation contributing a further 3 per cent to emissions (van der Werf et al. 2009). The Copenhagen Accord on Climate Change states:

> we recognise the crucial role of reducing emissions from deforestation and forest degradation and the need to enhance removals of greenhouse gas emissions by forests and agree on the need to provide positive incentives to such actions through the immediate establishment of a mechanism including REDD-plus, to enable the mobilization of financial resources from developed countries.[3]

The need to reduce emissions from the forest sector had also been stressed by the Intergovernmental Panel on Climate Change (2007) and the Stern Review (2006), which noted that reducing deforestation appeared to be a cost-effective approach to mitigating greenhouse gas emissions. Although implementing REDD may require more funds than initially estimated by the Stern Review, it still appears to be a cost-effective approach to emissions reduction (e.g. Boucher 2008; Kindermann et al. 2008).

There are a number of causes of deforestation and forest degradation (Geist and Lambin 2002), but fundamentally they occur because those who degrade and convert forests benefit from those activities. The benefits may be financial, for example, from higher returns generated by oil palm plantations compared with sustainable logging, or simply subsistence benefits, for instance, through the conversion of forest to crops for domestic consumption. Deforestation and degradation can involve actors from the local, national and even international scales, who benefit from forest transformation in different ways.

Funding is therefore needed to offset the forgone benefits of deforestation and degradation in developing countries, and to cover the costs incurred in implementing REDD policies and measures. The forgone benefits relate to the net economic benefits that would be derived, for example, from the conversion of a forest to an agricultural land use such as soy or oil palm cultivation. It is beyond the scope of this book to discuss various options for the design of a REDD scheme that are being considered by the Parties to the UNFCCC – for details of these proposals see Parker et al. (2008), Angelsen (2009) and Angelsen et al. (2009). It is useful, however, to outline some basic principles underpinning the design of a possible REDD scheme.

Essentially, REDD would involve: (1) setting a baseline level of emissions for a country; (2) that country implementing policies and measures to reduce emissions from deforestation and forest degradation during an agreed time frame (the so-called commitment period, expected to be about 5 years); (3) at the end of the commitment period, assessing emissions from forestry; (4) if a reduction in emissions below the baseline is achieved, issuing carbon credits[4] to the country and possibly to sub-national and non-State entities (if a market mechanism is included in an agreement on REDD); and (5) selling carbon credits through a financial mechanism.

One of the proposed approaches to setting the emissions baseline for a country involves estimating the emissions from forestry that occurred in that country over a certain past period (for instance, 1990 to 2000). This can be seen as the business-as-usual scenario, that is, the amount of forestry emissions that would occur without a REDD mechanism. Setting the baseline and measuring emissions at the end of the commitment period is necessary to assess whether REDD policies and measures can be assumed to have resulted in additional reductions of emissions.[5] That is, carbon credits can be considered as representing real avoided emissions. Parties to the UNFCCC seem to favour national level baselines over local level baselines in order to reduce the risk of displacement of greenhouse gas emissions from one geographic area to another, referred to as 'leakage'. For example, this may arise if clearing for agriculture is precluded in one area and, as a result, forest clearing takes place in a different area, whether carried out either by the same stakeholders or other parties that take advantage of the market opportunity, thus partly or completely offsetting the reduction in emissions.[6]

REDD policies refer, for example, to the reduction of incentives that lead to deforestation (for example, agricultural subsidies) and degradation (for instance, subsidized logging), adoption and enforcement of land use plans, and building state capacity to counter illegal forest activities. REDD measures refer to on-the-ground forest conservation activities (for example, the establishment of protected areas) that can be carried out by government, companies and local communities.

Funding for REDD could take three forms, not necessarily mutually exclusive, as noted by Angelsen et al. (2009). The first type of mechanism is a fund to be created with contributions from developed countries that commit to supporting the reduction of forestry emissions in developing countries. This fund could be used to build capacity for the implementation of REDD in individual countries and to prepare them for a market-based approach. The second type of mechanism is funding that is partially linked to the market: the carbon credits from REDD obtained by developing countries could be sold at auctions to interested parties, which could include countries and private companies, but these REDD carbon credits would not be exchangeable on carbon markets (that is, not fungible), such as the European Union Emissions Trading Scheme. The third type of mechanism involves REDD carbon credits being sold on carbon markets and being fully fungible with other carbon credits.

There is currently limited support for the third type of mechanism because of concerns that REDD carbon credits could flood the market as well as concerns about the permanence of REDD carbon credits. Market flooding could occur if a REDD mechanism generated a large supply of REDD carbon credits that would lower the average price of all carbon credits, thus resulting in lower incentives to reduce carbon emissions in developed countries. Permanence refers to whether the reduced emissions for which the REDD credits are issued will be reversed in the future, that is, whether a developing country that receives REDD carbon credits will experience emissions above the baseline in future commitment periods. At a local level, this can be visualized as a forest area being set aside to seques-ter carbon in one commitment period, but being cleared in the following one, thus resulting in emissions. Proposals for dealing with this issue include making REDD credits temporary, establishing REDD carbon credit reserves at the national level (for example, a certain percentage of the credits would not be sold), and making developing countries liable for the REDD carbon credits in case of non-permanence.

To implement REDD policies and measures within countries effec-tively and sustainably, there seems to be a need to link national with sub-national initiatives (Angelsen et al. 2008), which would involve the distribution of (some of) the revenues from the sale of REDD credits. The crucial question, therefore, is whether governments should provide incen-tives to individuals, communities and companies to reduce emissions from deforestation and forest degradation. It could be expected that govern-ments would provide incentives if reductions in emissions took place on private or community lands,[7] linked to the amount of carbon conserved by these stakeholders, possibly using a PES mechanism.

The situation is less clear for measures that address deforestation on government land, particularly in those countries where ownership is con-tested by local communities, and in countries where some publicly owned forest is allocated for community use or management (Table 1.1). The lack

Table 1.1 *Land tenure and governance in 20 countries with highest deforestation rates*

Country	Deforestation 2000–05 km²[a]	Contribution to global deforestation 2000–05 (%)	Ownership of forest (%)[b]				Freedom index 2000–05[c]
			Public, government managed	Public, community managed	Community ownership	Individuals and firms	
Brazil	155 150	24.1	21.0	6.1	25.9	47.0	F
Indonesia	93 570	14.5	98.4	0.2	0.0	1.4	PF
Sudan	29 450	4.6	95.8	4.2	0.0	0.1	NF
Myanmar	23 320	3.6	99.9	0.1	0.0	0.0	NF
Zambia	22 240	3.4	99.8	0.2	0.0	0.0	NF
Tanzania	20 610	3.2	89.6	4.5	5.8	0.2	PF
Nigeria	20 480	3.2	na	na	na	na	PF
Congo Dem. Rep.	15 970	2.5	100.0	0.0	0.0	0.0	NF
Zimbabwe	15 650	2.4	na	na	na	na	NF
Venezuela	14 380	2.2	100.0	0.0	0.0	0.0	PF
Bolivia	13 510	2.1	43.5	37.2	17.2	2.1	F/PF
Mexico	13 020	2.0	5.0	0.0	79.9	15.1	F
Cameroon	11 000	1.7	94.6	5.4	0.0	0.0	NF
Cambodia	10 940	1.7	100.0	0.0	0.0	0.0	NF
Ecuador#	9 880	1.5	77.1		22.9		PF
Australia	9 670	1.5	74.2	0.0	14.1	11.7	F
Paraguay	8 930	1.4	na	na	na	na	PF
Philippines#	7 870	1.2	89.5		10.5		F
Honduras#	7 820	1.2	75.0		25.0		PF
Argentina	7 490	1.2	20.4	0.0	0.0	79.6	F

Notes:

(a) Food and Agricultural Organization of the United Nations (FAO) (2006)
(b) Data from Sunderlin et al. (2008) for year 2008, except for: Mexico with data for 2002, and countries marked with #data from FAO (2006)
(c) Freedom House (www.freedomhouse.org)
NF: not free
PF: partly free
F: free; reported ranking indicates the dominant governance classification during the period.

of clarity about local entitlements to benefit from REDD schemes in countries with poor governance has led community and indigenous advocacy organizations to express concern about REDD. Griffiths (2007) states that the implementation of REDD schemes without due regard to rights, social and livelihood issues could increase the risks of:

- renewed and even increased state and 'expert' control over forests to protect lucrative forest carbon reservoirs;
- unjust targeting of indigenous people as the drivers of deforestation;
- violations of customary land and territorial rights;
- zoning of forest lands without the informed participation of forest dwellers by the state and/or non-government organizations;
- unequal imposition of the costs of forest protection on indigenous peoples and local communities;
- unequal and abusive community contracts;
- land speculation, land grabbing and land conflicts;
- corruption and embezzlement of international funds by national elites; and
- increasing inequality and potential conflict between recipients and non-recipients of funds.

Some of these had already been noted in the Stern Review (2006), which highlighted the risks of perverse incentives created through incorrectly set baselines, rent-seeking behaviour, the capture of benefits by national elites and corruption. It also stressed the need to clarify land and forest boundaries and ownership, and ensure any allocation of property rights is regarded as just by local communities (see also Cotula and Mayers 2009).

The foregoing concerns highlight the need to consider the impacts of REDD activities on individuals and communities, even when those activities might take place on state land. Apart from understanding social risks, this is important because if individuals and communities contribute to reducing emissions from deforestation and degradation on state lands, consideration should be given to rewarding them for their efforts. PES mechanisms could be used for that purpose.

PAYMENTS FOR ENVIRONMENTAL SERVICES AND LIVELIHOODS

A PES system is designed to provide payments to those who contribute to the provision of environmental services. Payments are aimed at changing resource use practices from those with negative economic impacts for society or for the 'buyer' (the party that purchases the environmental service) to those that are perceived to be beneficial. From an economic efficiency perspective, to make the implementation of a PES scheme worthwhile for society or the buyer, the value of the payment must be equal to, or lower than, the value of the avoided environmental change – the value of the environmental service provided. To be attractive to the seller, the value of the payment should be higher than (or at least equal to) her opportunity cost – the value of the benefits derived from the activities that would have been undertaken in the absence of the PES system.

A PES system is commonly defined as involving:

1. a voluntary transaction where
2. a well-defined ES [environmental service] (or a land-use likely to secure that service)
3. is being 'bought' by a (minimum one) ES buyer
4. from a (minimum one) ES provider
5. if and only if the ES provider secures ES provision (conditionality) (Wunder 2005, p. 3).

This definition is relevant to a market with active demand for, and supply of, an environmental service (Corbera et al. 2007a; 2007b). In practice, many PES schemes have been implemented in situations where there is no functioning market for these environmental services; and the existing literature uses the term 'market-based instruments' to refer to PES schemes across a spectrum from ideal market situations to those involving a hybrid of regulatory approaches with financial incentives, as well as development initiatives with financial incentives (Pagiola et al. 2002).

It is worth noting that Wunder's definition underplays the role of intermediaries in the transactions (Vatn 2010). Because environmental services are normally ill-defined and have the characteristics of public goods, they often require the intervention of an intermediary to regulate use, establish prices that are not set by markets and coordinate transactions between buyers and sellers. Basing PES only on buyer–seller interactions without recognizing the role of intermediaries restricts the application of the concept to a few peculiar cases, because most of the real-world situations do not in fact conform to the definition (Vatn 2010).While the foregoing definition of PES does not sufficiently stress the role of intermediaries, it needs to be recognized that the authors who have developed and used it acknowledge that many PES schemes have been financed by governments or non-government organizations who act as the intermediary for end users of the environmental service (Engel et al. 2008)

The research reported in this book worked with the definition outlined above in order to have an agreed benchmark for the case studies, primarily because it is widely accepted, and there was a need to maintain consistency across case studies. The problems arising from the strict adoption of this definition, however, cannot be discounted and are commented upon in the concluding chapter, which compares the case studies and draws lessons for the design of PES schemes for REDD activities.

Although PES was not developed as an instrument to improve liveli-hoods, there has been an increasing interest in the possible impacts of PES on participants (that is, sellers) and non-participants, particularly in rela-tion to their impacts on poverty (e.g. Grieg-Gran et al. 2005; Pagiola et al. 2005; Porras et al. 2008), an interest that has grown further as a result of the concerns about the potential impacts of REDD on livelihoods and rights over land and resources.

We use the term livelihoods to refer broadly to 'the capabilities, assets (stores, resources, claims and access) and activities required for a means of living' (Chambers and Conway 1991). Livelihoods are dynamic, involving continuous management and modification of assets, as well as choices to trade off and draw down on natural, financial, human, physical and social-political assets (Scoones 1998; Bebbington 1999). A livelihoods approach has informed some key analyses of the impacts of PES (Landell-Mills and Porras 2002; Miranda et al. 2003; Grieg-Gran et al. 2005), which have examined how PES interventions have interacted with different livelihood assets over time. Recent research has also emphasized issues of access and power in PES schemes, so that impacts are considered within a wider social and political context (Corbera et al. 2007a; Pagiola et al. 2008; Wunder 2008), which was a gap in early livelihoods research (de Haan and Zoomers 2005).

Much of the PES and livelihoods research has focused on impacts on 'the poor' (typically defined according to national poverty line bench-marks), in part because of the strong overlap between the resources and environmental services targeted by PES schemes and areas of high poverty incidence.

However, this overlap does not necessarily equate with high rates of participation by poor households in PES schemes, because of a number of 'participation filters' (Wunder 2008). The poor may be ineligible to participate for a number of reasons, including weak or unrecognized resource rights, because of a lack of awareness, or if the perceived costs (for instance, investment or opportunity costs) relative to the benefits of participation are too high (Pagiola et al. 2008; Wunder 2008). Yet despite their non-participation, these households may still experience impacts from PES schemes and therefore need to be considered when analyzing impacts (Wunder 2008).

The idea that the poor are, in fact, a diverse group has been recognized in some PES research (Pagiola et al. 2005; Corbera et al. 2007a; Wunder 2008), though the evidence on how impacts vary according to people's dif-ferent resources and capacities remains limited. Lee and Mahanty (2009), drawing on experiences in community-based forest management (CBFM) suggest that, without active intervention, the poorest households are less likely to benefit from CBFM schemes than those those with a minimal level of assets and resource security. Corbera et al. (2007a) add that the equity of PES impacts depends upon the social context. The existence of collective resource ownership with strong institutions has supported collective welfare improvements, but without these conditions the landless and poor can be excluded (Corbera et al. 2007b). For those concerned with distribution and equity, the relationship between PES, livelihoods and different types of poverty needs to be better understood.

Once households do gain access to a PES scheme, the question remains of how much and in what way they may benefit from participation, in either financial or non-financial terms (Wunder 2008). Analyses applying a livelihoods framework have made the useful contribution of directing attention away from a focus on income, to consider how PES schemes interact with a wider set of assets including natural, financial, human, physical and social-political assets. Given the diversity of PES schemes, as well as the environmental service sellers and buyers associated with them, a range of risks and opportunities can be shown to exist across the full spectrum of livelihood assets (Table 1.2).

A number of critical factors have been identified that influence whether the risks or opportunities outlined above are realized in practice, including (Landell-Mills and Porras 2003; Pagiola et al. 2005; Corbera et al. 2007a; Wunder 2008; Pagiola 2008):

- the nature and location of the environmental service, for instance, the percentage of poor households tends to be higher in remote areas where forests are often located;
- whether people have the recognized and secure resource rights generally needed to enable entry into PES agreements;
- whether workable regulatory frameworks exist for a specific environmental service;
- how many PES participants are poor, and their ability to participate;
- the size of the payment provided for the provision of the environmental service;
- finance and credit availability for sellers to cover their up-front costs of participation;
- the skills, education, power and negotiating capacity of environmental service sellers;
- availability of good market information and linkages related to communication infrastructure; and
- the existence of mechanisms to reduce transaction costs, for example, collective action institutions that facilitate coordination amongst environmental service sellers.

Table 1.2 Livelihood risks and opportunities in PES

Assets	Risks	Opportunities
Financial	• PES income may be concentrated amongst those who are able to participate • Benefits delivered at the community level may be distributed inequitably • Restrictions on agricultural expansion and resource use may reduce income from these sources • Potential increases in cost of living because of increased income to PES participants and increased land value • Reduced income for particular social groups (for example, women, informal settlers) if restrictions to marketable non-timber forest products apply	• Additional income to participating households • PES income may be relatively stable provided the scheme continues • Improved access to cash for investing in rejuvenating marginal/low productivity lands
Human	• Limited opportunities for the poor to capture capacity development opportunities • Reduced health if loss of access to non-timber forest products for direct use	• Many examples of education and training associated with PES initiatives • Improved health through improvements to water supply and air quality • Education and health could improve if PES income is invested in these

	Costs	Benefits

Natural	• Access to common lands for grazing, resource collection and swidden agriculture by marginal groups may be restricted
	• Increase in value of currently marginal land may increase incentive for powerful groups to take control of it
		• Strengthened tenure security in some cases:
		  • Land under PES agreement is not 'idle' and therefore prevents encroachment
		  • Tenure security used as a reward for ES provision
		  • Improvement in the status/value of natural resources

Social and political	• If collective decision-making processes are weak (lack of transparency, accountability), costs and benefits of PES may be distributed inequitably • Erosion of social cooperation if conflict amongst participants or between participants and non-participants • Cultural impact of monetizing ES	• Strengthening/creating institutions to negotiate agreements can build or strengthen social capital in communities • Greater visibility and ability to attract funds for some activities • Protection of natural and cultural heritage improves recreation and cultural opportunities • Potential to incorporate and 'certify' traditional (sustainable) forms of production
Physical	• Dismantling of local infrastructure for example, roads, to secure environmental services • Inequality in infrastructure development so that only market participants benefit	• Infrastructure development with community level payments/rewards – transport, market infrastructure, research, health care, housing, water supply, communications

Source: Landell-Mills and Porras 2002; Miranda et al. 2003; Grieg-Gran et al. 2005; Corbera et al. 2007a; Iftikhar et al. 2007; Scherr et al. 2007; van Noordwijk et al. 2007; Pagiola et al. 2008; van Noordwijk et al. 2008; Wunder 2008.

The empirical evidence regarding the impacts of PES on livelihoods is still limited (Wunder et al. 2008), but it shows the following. PES schemes appear to have generated small gains above the opportunity costs faced by participants (Wunder et al. 2008), gains which, given the nature of the schemes, normally involve increased income. The poor have been able to participate in PES schemes, even when they were not directly targeted (Grieg-Gran et al. 2005; Asquith et al. 2008; Engel and Palmer 2008), but high transaction costs are a significant threat to their participation (Grieg-Gran et al. 2005; Pagiola et al. 2008). Participants' non-income benefits include increased tenure security (Grieg-Gran et al. 2005; Asquith et al. 2008; Engel and Palmer 2008) and increased social capital. Some indirect negative impacts may also occur, such as reduced quality of roads (due to increased transport of produce) and of water (due to establish-ment of timber plantations) (Grieg-Gran et al. 2005). Limited knowledge exists about impacts on non-participants who – depending on the type of scheme and the changes it brings about – could benefit (for example, from improved water quality from watershed schemes) (Wunder 2008), or be negatively affected (for example, by lower labour demand and higher food prices possibly brought about by forest conservation PES schemes) (Grieg-Gran et al. 2005).

In summary, the existing body of knowledge highlights that, although PES raises a range of risks and opportunities for local livelihoods, the evi-dence on how these play out in practice is limited. This weak evidence base is the starting point for our research, as is the desire to understand better the key factors that might determine livelihood impacts. The analytical framework discussed below takes up these issues and sets out how they are dealt with in the case studies.

METHODOLOGY

This research applied a comparative case study method. Nine PES schemes from a range of countries across Africa, Asia and Latin America were selected, with preference given to those from the 20 countries with the highest annual deforestation (Table 1.1), which accounted for approxi-mately 80 per cent of total deforestation between 2000 and 2005. As not all case studies could be selected from this group, cases were chosen to represent as closely as possible the social and economic conditions of those countries that make the highest contributions to deforestation.

Case studies were not restricted to focusing on PES schemes that aimed to reduce deforestation or sequester carbon (though some of them do so), but were chosen to represent a range of different environmental services, including biodiversity conservation and watershed management. The selected case studies had preferably been implemented for long enough to allow the assessment of emerging livelihood impacts, and had to address the design features of the PES scheme that have led to positive or negative livelihood impacts. The selection of case studies was also based on a desire for them to represent a range of different implementation scales – from small-scale through to national schemes – as well as those for single services or multiple (bundled) services.

The authors of the case studies were asked to address, as far as possible, a common set of conditions and issues in their analysis which, as noted earlier, draws on the sustainable livelihoods framework, with particular attention to distributional and access issues. Accordingly, case studies start with some basic background and history on the PES scheme, the type of environmental service targeted, the background on environmental service buyers and sellers, and the role of any intermediaries and the government. The main design features of the PES scheme are outlined, including: the geographical coverage of the scheme; how the service is defined and measured/monitored, and who does that; how the price is set and by whom; how sellers are selected and their socioeconomic characteristics; whether the payments are in cash and/or in-kind; and the roles and responsibilities of buyers, sellers and intermediaries. Also discussed are the terms of the agreements (time period, fixed-term or on-going, etc.), and periodicity of payments; any sanctions for non-compliance by sellers and buyers; the transaction costs (including costs of contracting individual participants and the costs on the participants); and any technical barriers to participation as well as technical support necessary to overcome these.

The impacts of the PES scheme were analysed in relation to the biophysical environment, the livelihoods of participants, the livelihoods of non-participants, and other factors or processes beyond the PES scheme that might have had an influence. A number of specific questions were identified for each livelihood asset category (Table 1.3) and authors were asked to consider, as far as possible, different impacts on different wealth strata, for instance, the poorest versus less poor.

Individual case studies were not expected to address all of these questions, given the diversity of schemes assessed, their associated impacts and the level of available data. However, by articulating a common set of questions, we invited authors to go through a systematic process of considering which livelihood assets were affected in their case study, and how.

Table 1.3   Questions on livelihood impacts of PES schemes

Assets	Key questions
Financial	Does the PES scheme increase the overall income of participating households? (compared with opportunity costs of alternative activities, appropriately discounted)
	Is a diversity of income sources for participants sustained?
	Does the PES scheme contribute to increases in the cost of living?
Human	Does the PES scheme improve capacity, skills and knowledge, and for whom?
	Does the PES scheme impact on health?
	Is PES income (especially at the community level, if any) invested in education and health improvements?
Natural	Does the PES scheme contribute to a change in access to resources, particularly in common property regimes?
	Does the PES scheme result in a change of the perceived status/value of natural resources?
	Does the PES scheme affect resource tenure (that is, land tenure, access to common resources)?
	Does the PES scheme affect cultural motivations for environmental protection?
Social/political	Does the PES scheme impact on the social capital of the relevant local communities?
	Does the PES scheme impact on coordination and influence with wider institutions and decision-making processes?
Physical	Does the PES scheme impact on investment in local infrastructure?

Sources: Adapted from Chambers and Conway 1991; Scoones 1998; Ritchie et al. 2000; Landell-Mills and Porras 2002; Miranda et al. 2003; Grieg-Gran et al. 2005; Grieg-Gran et al. 2006; Corbera et al. 2007a; Iftikhar et al. 2007; van Noordwijk et al. 2007; Pagiola et al. 2008; Wunder 2008; Lee and Mahanty 2009.

THE STRUCTURE OF THE BOOK AND THE CASE STUDIES

The following chapters present the nine case studies (see Table 1.4), while the overall findings from the comparative analysis are presented in the concluding chapter. Five of the case studies are in countries listed in the top 20 deforesting countries (Table 1.1); three are in geographic areas that have a country in the top 20 list, and one study is a review of PES projects funded by the Global Environment Facility (GEF). The latter was not in

Table 1.4   Characteristics of case studies

Country	Environmental service targeted	Scale of scheme	Ownership of land
Global GEF review	Variety of case studies including carbon, watershed protection	National to sub-national	State, common property, private
Mexico*	Carbon (also hydrological services and biodiversity)	National	Common property
Brazil*	Bundle – reduced deforestation, carbon sequestration, biodiversity conservation, hydrological functions, fire management	Sub-national (nine states of the Amazon region)	Private
Indonesia*	Watershed protection	Sub-national (watershed spanning two regencies and six sub-districts)	Private
Philippines*	Watershed protection	Sub-national (spanning multiple local government units)	State
Uganda	Carbon	Sub-national (small scale)	Private and state
Mozambique	Carbon	Sub-national (small scale)	Common property
Nicaragua and Colombia	Biodiversity conservation, carbon sequestration	Sub-national (small scale)	Private

Notes: * Country ranked amongst the top 20 deforesting countries (as listed in Table 1.1).

the original design but it was selected because it provided an opportunity to consider more than one PES scheme in one chapter.

Chapter 2 by Haskett and Gutman reviews 22 PES-related projects that have received funding from the GEF since 2001 – approximately US$170 million had been invested in PES and PES-related projects by 2008, with further investments made by GEF partners. The bulk of this portfolio is located in Latin America, with the remainder in Africa – no PES schemes are funded by the GEF in Asia – and most projects are relatively small-scale, although a significant number are of national scale, as well as two multinational projects and a global one. Virtually all GEF-supported PES projects aim to protect biodiversity; however, watershed management, carbon sequestration and landscape conservation for tourism are the activities that trigger payments. With the exception of the Costa Rican and Mexican national government-led PES schemes, all GEF-supported PES projects expect that the demand for ecosystem services will come, in the long term, from private buyers. However, at present, buyers are mostly national governments and donor agencies, with private buyers a distant third. Ecosystem service providers in these projects are predominantly small farmers, with only few medium and large landowners and protected area agencies as major sellers.

Haskett and Gutman go on to describe two projects in Costa Rica and Mexico in more detail. The first, in Costa Rica, is a national scheme to reduce deforestation and increase reforestation which would subsequently increase the supply of carbon sequestration, maintain forest hydrological services, biodiversity conservation, and preserve scenic beauty. Participants in this project believed that PES payments had made a significant improvement in income diversification and stabilization, and had offset start-up costs. The project is thought to have had positive environmental benefits, but has inadequately targeted the most vulnerable habitats. The Mexican scheme is a national scheme focusing on increasing fresh water availability, particularly groundwater, while having positive socioeconomic impacts on marginalized communities (particularly *ejidos* – rural communities on communal lands) (see also Chapter 3 by Corbera). The majority of PES payments are made to *ejidos* rather than individual landholders, though some *ejidos* have subsequently distributed income to the household level. There is evidence of strong tendencies toward equitable distribution of benefits within the communities (for example, school improvement). However, this programme appears to have been more successful in achieving social rather than environmental objectives, with overexploited aquifers under-represented in the programme.

Mexico's PES scheme for carbon payments was established in 2004 as part of a wider PES programme which covered carbon, biodiversity and agroforestry, and watershed services, and is described by Corbera in Chapter 3. In 2006, all the schemes were merged into a single policy framework, though each sub-programme maintained its own rules and procedures. The programme was implemented on private, state and community land, and involved an initial application for funding for project design, with a requirement to prove land ownership within a deforested or degraded area of the country. Proposals were then independently assessed before successful applications proceeded to implementation. Despite

application procedures and eligibility criteria for carbon projects changing several times over the life of the programme, overall the group of PES
programmes have been successful in attracting a large number of rural
communities and individual landholders – over 2600 communities, associations and private right-holders on more than 1.75 million ha were receiving payments for various ecosystem services by 2008. Corbera describes
how uncertain funding arrangements have meant that some approved
applications have not received funding to implement their projects, and
how the high rejection rates of applications highlight the need for sufficient
capacity amongst both communities and intermediaries, and means of
building this knowledge and capacity over time.

In terms of livelihood impacts, Corbera provides evidence that projects
contribute to increased awareness of conservation and ecosystem services,
and to strengthened forest management skills amongst participants, though
it was felt that further capacity-building efforts were needed. Positive contributions to human capital have also been felt by intermediary organizations, who have received training to improve the assistance provided to
communities. Impacts on physical capital have been uneven, depending
on the way community-level income has been spent. While the primary
reason for joining the PES programme was to earn additional income, the
majority of participants felt that contributions to financial capital were
insufficient to make a significant contribution to livelihoods in most cases.
However, contributions to social capital have been made through community organization, and strengthening collective action through land use
management and tree planting activities. On rare occasions, project activities have resulted in conflict within communities. The relatively high rates
of participation by marginalized communities is thought to be a highlight
of the programme, and the majority of participants have demonstrated
a willingness to conserve land regardless of the level of PES incentives,
reinforcing positive attitudes towards conservation activities.

In Chapter 4, Bartels et al. describe *Proambiente*, a programme implemented across nine states in the Amazon region of Brazil, though the
chapter focuses on sites within Acre state. The programme provides
(financial) rewards to smallholders for developing and implementing long-
term, sustainable land management plans – over land that can be used for
cultivation, livestock rearing and/or forest extractive activities – which is
assumed to deliver a bundle of environmental services including reduced
deforestation, carbon sequestration, biodiversity conservation, the restoration of hydrological functions and a reduction in uncontrolled use
of fire. Additional benefits provided to smallholders by the programme
include technical assistance, credit and participation in local and regional
planning processes – it was never intended that the payments would

provide the main incentive for family participation. Starting in 2003, *Proambiente* was designed to be implemented over 15 years, in a phased approach starting with the creation of regional sustainable development plans, then the development of management plans for each family, with these plans subsequently implemented. The establishment of community agreements and certification plans, audit and certification of activities and payment for environmental services would then follow.

In terms of impacts on households, Bartels et al. describe how the programme has affected financial capital minimally (though indirect impacts may have been quite strong – in terms of linking producers with specialized markets), the technical capacity of smallholders (and extension agents) has been strengthened, and the use of use of community agents (that is, local farmers trained to become trainers) has helped build both human and social capital. Some of the introduced practices have raised farm labour requirements (as yet uncompensated), and there remains some confusion about them. Social capital has also been built through the provision of information about, and the development of partnerships with, other organizations. The development of the programme and influence at the state and federal level have raised political capital, and have helped lay the groundwork for the establishment of true PES programmes in Brazil.

Indirect positive environmental impacts of *Proambiente* include the reforestation of riparian zones, and perhaps most prominently, the success in promoting controlled-burn techniques and reducing the use of fire for shifting agriculture. However, the fragmented nature of the smallholdings means that the programme cannot guarantee the provision of environmental services (for example, where fire from an adjacent non-participating property escapes onto *Proambiente* farms).

In Chapter 5, Leimona et al. describe an Indonesian PES scheme in the Cidanau watershed, one of the most important watersheds for supplying domestic and industrial water needs of Banten Province, Java. The watershed also has a biodiversity protection role, containing the only remaining lowland swamp forest in Java. Participating villages were selected based on mapping of critical areas (for example, steep slopes and erosion-prone soil) and participating farmers in each village were selected on the basis of their private land ownership and involvement in farmer groups. Contracts were offered to farmer groups to implement forest rehabilitation activities. Each group member was responsible for his or her own area, but the members were jointly responsible for meeting group commitments. Leimona et al. describe how the scheme had only one private-sector buyer of environmental services, with a multi-stakeholder forum playing an intermediary role between buyers and sellers in managing PES payments; supporting activities on relevant farms; encouraging potential buyers to join the scheme; and liaising between provincial and district government levels. The small scale of the PES scheme means that it may have limited positive environmental impacts – it covers only 100 ha of a 22 260 ha watershed – and there are uncertain relationships between the land use practices and water quality and quantity in the watershed.

In terms of livelihood impacts, Leimona et al. note that while there were no significant changes in income sources resulting from the PES scheme, the tree species planted were selected based on commodity prices and market demand to enable participants to build their production base for valuable tree crops, though some participants lost income-earning opportunities from activities no longer allowed under the scheme. One important impact of the PES scheme has been the stimulation of local business as an alternative to the agricultural sector, mostly because of additional business development support from government and non-government agencies involved in the PES scheme. Human capital had been built, with improved knowledge of environmental issues, although confusion about the details of PES contracts remained amongst a majority of participants. Leimona et al. also describe improvements in social capital, and the investments in physical capital that have been made in some villages, based on collective decisions taken by participants (and sometimes also with non-participants).

Soriaga and Annawi in Chapter 6 analyse the 'No-Fire Bonus' (NFB) scheme, launched in 1996 in the fire-prone pine forests of Mountain Province, Cordillera Administrative Region in the northern Philippines. While the scheme involved payments for fire protection, it was not developed with attention to PES principles, but did aim to contribute to national watershed management objectives by reducing uncontrolled fires, contributing to rainwater infiltration, soil fertility, natural regeneration and enhancing wildlife habitats. The scheme was a break from the command-and-control methods of the past, offering financial incentives to change the behaviour of those residing on the land within the scheme, much of which fell under the ancestral domain of several indigenous groups.

The scheme ran for only one monitoring and payment cycle, between 1996 and 1998, but reportedly gained wide recognition, appears to have had long-term social and environmental impacts well beyond its funding cycle, and offers important lessons about the role of government actors and about sustainability issues in such schemes. The project is a good example of the complexities resulting from informal arrangements governing such programmes – the agreement between the government department involved the Congressman and Governor of the province (who had agreed to be the buyers of the ES), was never formalized, and one of the buyers subsequently failed to deliver payments. The sellers in the NFB scheme were *barangay* (village) governments on behalf of their residents, who controlled their use of fire in return for (funding for) an infrastructure project, as chosen by the *barangay* government.

The most immediate and direct impact of the NFB scheme was the financial benefit obtained by households that provided labour, materials and services during construction of the infrastructure projects such as drainage canals, erosion prevention works and water tanks. The drainage canals constructed have remained operational and in use 10 years later. The sense of ownership of the project was gained in large part from efforts to control fire and from their choice of infrastructure project, which has subsequently led to better maintenance of that infrastructure. The boost to traditional methods of fire prevention was also of significant benefit, as was the social capital built between *barangay* governments and their residents.

In Chapter 7, German et al. present a case study on the Trees for Global Benefits Programme in Bushenyi District, Uganda, which aims to develop and implement a carbon trading model which could be replicated in other parts of the country, by paying smallholders for carbon credits associated with tree planting.

Participants are individual farmers or farmer groups that own, or have user rights over, a minimum area of land (often leases over private or state land) who are contracted to grow trees on their land for 25 or 50 years. Since its establishment in 2003, the programme has grown to involve around 200 farmers, with more applications in process – additional ES sellers are not accepted until buyers have been found. Participants are selected on a first-come first-served basis, but only if they have tenure security (though this does not require a title deed[8]), while those living close to protected areas are given priority. Though contracts are signed with individual farmers, they then form groups in order to minimize the costs associated with training and monitoring. A number of intermediaries are involved in the programme – providing capacity building, administrative support, etc. – which receive almost 30 per cent of revenue generated. However, these funds are insufficient to meet all operating costs, so the sustainability of the project is questionable. Of the remaining revenue, the majority is paid to individual households, and a Carbon Community Fund has recently been established to act as a form of insurance for use by farmers in case of accidents.

German et al. describe the impacts of the programme as being variable. The financial capital of carbon sellers has increased as a result of the payments, the lump sum nature of which enables sizable purchases or investments, in addition to indirect benefits (for example, increased availability of firewood and fruit). Increases in human and social capital were also reported by participants, though considerable confusion regarding the details of the contracts remains. German et al. report some reduction in agricultural productivity as reported by participants, and the reduced availability under informal arrangements of agricultural land for non-participants.

The Nhambita Community Carbon Project is the subject of Chapter 8 by Jindal. It is located in Sofala Province, Mozambique, in the buffer zone of Gorongosa National Park. The project began in 2003 and pays local smallholders to undertake agroforestry for carbon sequestration on their farms (covering approximately 1000 ha) and to conserve an area of communally owned miombo woodland (in excess of 11 000 ha). The project also promotes micro-enterprise development and undertakes research, extension and capacity building. Farmers can choose the agroforestry activities that most suit them from a 'menu' including horticulture, planting woodlots and intercropping, and all households are expected to refrain from burning forested areas for new farmlands. Contracts are offered to households for one or more activities for 100 years, though all payments are made within the first 7 years, after which it is hoped that the benefits arising from the agroforestry systems will provide sufficient incentive to continue. Payments are also made to the community for the successful protection of an area of communal woodland from any kind of harvesting and fire. The project also developed and supports a number of micro-enterprises (for instance, plant nurseries, a community saw mill and carpentry shops).

The project has a high penetration rate, with about 80 per cent of the local households already enrolled in different agroforestry contracts. Direct financial benefits are felt either as payments for agroforestry contracts (much of which was spent on consumer goods, education expenses or investments in agriculture), and/or as wages for employment with micro-enterprises. Jindal's evidence suggests that households receiving wages appear to be better off, though the same does not yet apply to households that receive only carbon payments (which are insufficient to significantly affect household financial capital). Significant contributions to human capital are thought to be the result of the training provided to smallholders, as well as on-the-job training of micro-enterprise employees; however, the project may have inadvertently increased women's workload. The project worked with local institutions, enabling better coverage of remote members of the community whilst strengthening these institutions and contributing to an increase in social capital. Educational and other physical infrastructure have also witnessed an impressive growth in the area.

Rios and Pagiola describe the Regional Integrated Silvopastoral Ecosystem Management Project, which was implemented in Colombia, Nicaragua and Costa Rica from 2003 to 2008, though Chapter 9 deals only with the project as implemented at sites in Colombia and Nicaragua. This project paid participating farmers to adopt a range of silvopastoral practices in degraded pastures in order to deliver biodiversity conservation and carbon sequestration services. Indirect benefits of these practices are anticipated to be additional production and diversification (for example, fruit, fuelwood, fodder and timber) and the improvement and/or maintenance of pasture productivity. A series of indices was developed to measure the biodiversity conservation and carbon sequestration services provided, which were then aggregated into an 'environmental services index'; payments were based on the change in this index. Payments were made annually for 4 years, with a one-off payment at the beginning of the project (in recognition of the environmental services households already provided), starting in 2004, and continuing until 2008. Participants were selected on a first-come basis in each of the areas, until the available funding was exhausted.

Rather than taking a strict livelihoods approach to determining the household-level impacts of this programme, Rios and Pagiola focus on the extent to which poor households were able to participate in the programme – in order to determine whether it was possible for poorer households to receive any benefits delivered by the project. The authors describe the examination of factors likely to affect participation and how, at both sites, poorer households have accounted for a substantial share of land use changes, which were not limited to the adoption of simpler and cheaper practices, and how across a number of different measures, it appears that poorer households were able to participate quite extensively in the PES scheme.

The impacts of the PES schemes on livelihoods are compared in Chapter 10, which uses the foregoing findings to derive lessons for the design of REDD schemes.

NOTES

1. The term used for REDD in the UNFCCC is REDD-plus, which includes deforestation, forest degradation and enhancement of carbon stocks. In this book, REDD is used for simplicity to refer to REDD-plus.
2. Deforestation refers to human-induced conversion of forest to non-forest cover. Forest degradation refers to negative changes in carbon density of the forest.
3. http://unfccc.int/resource/docs/2009/cop15/eng/l07.pdf
4. Carbon credit is used here to refer to the certified units of greenhouse gases that have not been emitted as a result of REDD policies and measures, also referred to as REDD units; one unit being one ton of carbon dioxide equivalent.

5. It cannot be proven with certainty that the reduction in emissions is due solely to REDD policies and measures.
6. Concern about leakage to other countries has resulted in proposals that suggest setting global baselines (e.g. Mollicone et al. 2007).
7. This approach implies that reduced deforestation involve a geographically referenced approach that would earmark specific forest areas for carbon conservation.
8. Local authorities assist with the verification of ownership, through records of prior land purchase agreements and wills.

REFERENCES

Angelsen, A. (ed.) (2009), *Realising REDD+: National Strategy and Policy Options*, Bogor: Center for International Forestry Research.

Angelsen, A. (2008), *Moving Ahead with REDD: Issues, Options and Implications*, Bogor: Center for International Forestry Research.

Angelsen, A., S. Brown, C. Loisel, L. Peskett and C. Streck (2009), Reducing Emissions from Deforestation and forest Degradation (REDD): an options assessement report. Prepared for the Government of Norway. Washington DC: Meridian Institute.

Angelsen, A., C. Streck, L. Peskett, J. Brown and C. Luttrell (2008), What is the right scale for REDD? in A. Angelsen (ed.) *Moving Ahead with REDD: Issues, Options and Implications*, Bogor: Center for International Forestry Research.

Asquith, N.M., M.T. Vargas and S. Wunder (2008), Selling two environmental services: in-kind payments for bird habitat and watershed protection in Los Negros, Bolivia. *Ecological Economics*, **65** (4), 675–84.

Bebbington, A. (1999), Capitals and capabilities: a framework for analyzing peasant viability, rural livelihoods and poverty. *World Development*, **27** (12), 2021–44.

Boucher, D. (2008), *Out of the Woods: A Realistic Role for Tropical Forests in Curbing Global Warming*, Cambridge: Union of Concerned Scientists.

Chambers, R. and G. Conway (1991), *Sustainable Rural Livelihoods: Practical Concepts for the 21st Century*, Sussex: Institute of Development Studies.

Corbera, E., K. Brown and W.N. Adger (2007a), The equity and legitimacy of markets for ecosystem services. *Development and Change*, **38** (4), 587–613.

Corbera, E., N. Kosoy, and M. Martinez-Tuna (2007b), The equity implications of marketing ecosystem services in protected areas and rural communities: case studies from Meso-America. *Global Environmental Change*, **17** (3–4), 365–80.

Cotula, L. and J. Mayers (2009), 'Tenure in REDD: starting point or after-thought?' Natural Resource Issue No. 15. London: International Institute for Environment and Development.

de Haan, L. and A. Zoomers (2005), Exploring the frontier of livelihoods research. *Development and Change*, **36** (1), 27–47.

Engel, S., S. Pagiola and S. Wunder (2008), Designing payments for environmental services in theory and practice: an overview of the issues. *Ecological Economics*, **65** (4), 663–74.

Engel, S. and C. Palmer (2008), Payments for environmental services as an alternative to logging under weak property rights: the case of Indonesia. *Ecological Economics*, **65** (4), 799–809.

Food and Agriculture Organization of the United Nations (2006), *Global Forest Resources Assessment 2005: Progress Toward Sustainable Forest Management*, Rome: FAO.

Geist, H.J. and E.F. Lambin (2002), Proximate causes and underlying driving forces of tropical deforestation. *BioScience*, **52** (2), 143–50.

Grieg-Gran, M., S. Noel and I. Porras (2006), *Lessons Learned from Payments for Environmental Services*. Wageningen: ISRIC.

Grieg-Gran, M., I. Porras and S. Wunder (2005), How can market mechanisms for forest environmental services help the poor? Preliminary lessons from Latin America. *World Development*, **33** (9), 1511–27.

Griffiths, T. (2007), *Seeing 'RED'? 'Avoided Deforestation' and the Rights of Indigenous Peoples and Local Communities*, Moreton-in-Marsh: Forest Peoples Programme.

Iftikhar, U.A., M. Kallesoe, A. Duraiappah, G. Sriskanthan, S.V. Poats and B. Swallow (2007), *Exploring the Inter-linkages Among and Between Compensation and Rewards for Ecosystem Services (CRES) and Human Well-being*, Nairobi: World Agroforestry Centre.

Intergovernmental Panel on Climate Change (2007), Technical summary: contribution of Working Group III to the Fourth Assessment Report of the Intergovernmental Panel on Climate Change. Intergovernmental Panel on Climate Change.

Kindermann, G., M. Obersteiner, B. Sohngen, J. Sathaye, K. Andrasko, E. Rametsteiner, B. Schlamadinger, S. Wunder and R. Beach (2008), Global cost estimates of reducing carbon emissions through avoided deforestation. *Proceedings of the National Academy of Sciences*, **105** (30),10302–307.

Landell-Mills, N. and I. Porras (2002), *Silver Bullet or Fools' Gold? A Global Review of Markets for Forest Environmental Services and their Impact on the Poor*, London: International Institute for Environment and Development.

Lee, E. and S. Mahanty (2009), *Payments for Environmental Services and Poverty Reduction: Risks and Opportunities*. Bangkok: RECOFTC.

Miranda, M., I. Porras and M.L. Moreno (2003), *The Social Impacts of Payments for Environmental Services in Costa Rica. A Quantitative Field Survey and Analysis of the Virilla Watershed*, London: International Institute for Environment and Development.

Mollicone, D., F. Achard, S. Federici, H. Eva, G. Grassi, A. Belward, F. Raes, G. Seufert, H.-J. Stibig, G. Matteucci and E.-D. Schulze (2007), An incentive mechanism for reducing emissions from conversion of intact and non-intact forests. *Climatic Change*, **83** (4), 477–93.

Pagiola, S., A. Arcenas and G. Platais (2005), Can payments for environmental services help reduce poverty? An exploration of the issues and the evidence to date from Latin America. *World Development*, **33** (2), 237–53.

Pagiola, S., J. Bishop and N. Landell-Mills (2002), *Selling Forest Environmental Services: Market-Based Mechanisms for Conservation and Development*, London: Earthscan Publications.

Pagiola, S., A.R. Rios and A. Arcenas (2008), Can the poor participate in payments for environmental services? Lessons from the Silvopastoral Project in Nicaragua. *Environment and Development Economics*, **13** (3), 299–325.

Parker, C., A. Mitchell, M. Trivedi and N. Mardas (2008), *The Little REDD Book: A Guide to Governmental and Non-governmental Proposals to Reducing Emissions from Deforestation and Degradation*, Oxford: Global Canopy Programme.

Porras, I., M. Grieg-Gran and N. Neves (2008), *All that Glitters: A Review of Payments for Watershed Services in Developing Countries*, London: International Institute for Environment and Development.

Ritchie, B., C. McDougall, M. Haggith and N.B. de Oliviera (2000), *Criteria and Indicators of Sustainability in Community Managed Forest Landscapes*, CIFOR, Bogor: International Institute for Environment and Development.

Scherr, S., J.C. Milder and C. Bracer (2007), *How Important will Different Types of Compensation and Reward Mechanisms be in Shaping Poverty and Ecosystem Services Across Africa, Asia and Latin America over the Next Two Decades?* Nairobi: World Agroforestry Centre.

Scoones, I. (1998), *Sustainable Rural Livelihoods: A Framework for Analysis*, Sussex: Institute of Development Studies.

Stern, N. (2006), *Stern Review on the Economics of Climate Change*, London: UK Treasury/Cabinet Office.

Sunderlin, W.D., J. Hatcher and M. Liddle (2008), *From Exclusion to Ownership? Challenges and Opportunities in Advancing Forest Tenure Reform*, Washington DC: Rights and Resources Initiative.

van der Werf, G.R., D.C. Morton, R.S. DeFries, J.G.J. Olivier, P.S. Kasibhatla, R. B. Jackson, G.J. Collatz and J.T. Randerson (2009), CO_2 emissions from forest loss. *Nature Geoscience*, **2** (11), 737–38.

van Noordwijk, M., B. Leimona, L. Emerton, T.P. Tomich, S.J. Velarde, M. Kallesoe, M. Sekher and B. Swallow (2007), *Criteria and Indicators for Environmental Service Compensation and Reward Mechanisms: Realistic, Voluntary, Conditional and Pro-poor*, Nairobi: World Agroforestry Centre.

van Noordwijk, M., H. Purnomo, L. Peskett and B. Setiono (2008), *Reducing Emissions from Deforestation and Forest Degradation (REDD) in Indonesia: Options and Challenges for Fair and Efficient Payment Distribution Mechanisms*, Nairobi: World Agroforestry Centre.

Vatn, A. (2010), An institutional analysis of payments for environmental services. *Ecological Economics* **69** (6), 1245–52.

Wunder, S. (2008), Payments for environmental services and the poor: concepts and preliminary evidence. *Environment and Development Economics*, **13** (3), 279–97.

Wunder, S. (2005), *Payments for Environmental Services: Some Nuts and Bolts*, Bogor: Center for International Forestry Research.

Wunder, S., S. Engel and S. Pagiola (2008), Taking stock: a comparative analysis of payments for environmental services programs in developed and developing countries. *Ecological Economics*, **65** (4), 834–52.

2. Taking stock of the Global Environment Facility experience with payments for environmental services projects

Jonathan Haskett and Pablo Gutman

INTRODUCTION

This chapter differs from the others in this volume as it reviews not one, but a portfolio of 22 payments for environmental services (PES) related projects that received grants from the Global Environment Facility (GEF) between 2001 (when the first PES project appeared in the GEF portfolio) and late 2008. Although this portfolio gives little guidance about the design of an international regime for reduced emissions for deforestation and forest degradation (REDD), it provides many lessons about the design of national REDD schemes, which are important pieces of the REDD puzzle.[1]

The GEF is the major multilateral fund created to grant money for conservation activities in developing countries and economies in transition. Between 1991 and 2006 the GEF disbursed some US$7.4 billion which, with US$28 billion of co-financing, supported some 2000 projects in over 160 developing countries. Arguably the most important multilateral environmental agreements have designated the GEF to be their funding mechanism, among them, the United Nations Framework Convention on Climate Change (UNFCCC), the Convention on Biological Diversity, the United Nations Convention to Combat Desertification and the Stockholm Convention on Persistent Organic Pollutants. The GEF is also a major funder for projects related to the Montreal Protocol for the Phasing Out of Ozone Depleting Substances. If funding for the UNFCCC and other multilateral environmental agreements increases, part of the additional funding will most likely be channelled through the GEF.

The GEF has a small but significant portfolio of PES projects in developing countries, it is an important player in the global environmental arena in terms of its mandate, funds and partners, and its portfolio (for PES and REDD) is likely to grow. A majority of the PES projects in the GEF portfolio are already REDD-like projects – that is, they may not refer directly to REDD, but reducing deforestation is either their main objective or forest conservation is the vehicle to deliver other objectives such as watershed or biodiversity protection (see Table 2.1).

This chapter discusses the significance of the GEF and provides an overview of the GEF PES portfolio and key elements in the design of these projects. By late 2008, only three projects – the PES programmes of Costa Rica and Mexico, and the regional silvopastoral PES project – were advanced enough to render significant information on their implementation impacts. Detail is provided on the Costa Rica and Mexico programmes in order to learn more about their social and environmental design and impacts (see Chapter 9 for more information on the silvopastoral project). Finally, the findings of this review and some lessons for the design of pro-poor REDD projects are presented.

A Snapshot of the GEF Portfolio

All GEF PES projects began after 2000, and they are still a minor component of the GEF funding, with an investment of approximately US$170 million (and co-financing of US$700 million) in 22 PES or PES-related projects by mid-2008 (see Appendix 2.1 for a list of each of these projects). This represents less than 4 per cent of the GEF portfolio, as measured by number of projects and by money granted.[2] Part of the reason that an organization that conceptually embraces an ecosystem approach to environmental conservation has such a small PES portfolio is that the schemes are difficult to implement, and the short project cycle of GEF grants is ill-suited to the long-term payment commitment required for a PES scheme (Wunder and Wertz-Kanounnikoff 2008).

Measured by international standards, GEF PES projects are of medium or small size. The majority have total budgets of US$25–100 million, and there is only one greater than US$100 million. The portfolio is heavily concentrated in Latin America (73 per cent of the total),[3] with only five projects in Africa (23 per cent) and one global project (5 per cent). There are no projects in Asia.

Most projects are of local scale, but a significant number are of national scale and there are two multinational projects and a global one (see Table 2.1). The projects are classified into four categories – stand-alone large-scale projects; stand-alone small-scale projects; projects where PES is a small but distinctive component of a larger project; and projects where

Table 2.1 Summary of the GEF PES portfolio

Region	No. of projects	Geographical scope		PES focus				Ecosystem services targeted			
		National/ global	Local	Stand alone large PES	Stand alone small PES	Many objectives including an operational PES scheme	Many financial alternatives including PES	Biodiversity conservation	Watershed protection	Carbon sequestration	Landscape aesthetic values (for tourism)
Africa	5	–	5	–	1	1	3	4	4	1	1
Latin America	16	6	10	3	1	8	4	16	12	7	3
Global	1	1	–	1	–	–	–	1	1	1	–
Total	22	7	15	4	2	9	7	21	17	9	4

Region	Activities to achieve ecosystem services						
	Protect natural forests	Reforestation	Silvopastoral practices	Agroforestry practices	Other sustainable agriculture practices	Protected area management	Buffer zone management
Africa	2	1	1	2	3	2	1
Latin America	7	7	4	8	9	4	5
Global	1	1	1	1	1	–	–
Total	10	9	6	11	13	6	6

Source: GEF (2008) project database.

PES is a small and ill-defined component – which have implications for design, implementation and monitoring.

Stand-alone large-scale PES projects (for example, in Costa Rica and Mexico) are large projects covering hundreds of thousands of hectares and with thousands of participants. They are government-driven (that is, government is the largest source of funding), and their design included detailed consideration of all major PES questions (for instance, what were the environmental services, who were the buyers and sellers, what prices will be used, etc.). The GEF added modest funds to these projects (around 10 per cent of the total budgets) and focused on improving the PES design and targeting, though these initial designs (for example, Costa Rica eco-markets) gave limited attention to measuring and monitoring environmental and social impacts.

Stand-alone small-scale PES projects (for instance, the silvopastoral project and the Uganda Experimental PES) range in size from hundreds to a few thousand hectares, and from dozens to a few hundred participants. They are driven and funded by the GEF and its implementing agencies and their purpose is mainly learning and demonstration, and thus they have been carefully designed and closely monitored.

In many cases where PES is a small but distinctive component of a larger agricultural development, sustainable agriculture or conservation project (for example, the Kenya Agricultural Productivity Project, the Dominican Republic Demonstrative Sustainable Land Management Project), original project documents say little about the PES component, and leave the design of the PES until the project implementation phase. In all of these projects, the PES component is an addition to a basket of funding mechanisms that may include increased rural productivity, increased market access and certification of organic or environment-friendly products.

Where PES is a small and ill-defined component (for example, the South Africa National Grassland Initiative, the Colombia National Protected Areas Conservation Trust Fund, the Peru Strengthening Biodiversity Conservation Project), the initial project design often only commits to the exploration of PES as one of several financing alternatives, so the completed project may or may not have a PES component. However, the GEF Secretariat considers them to be part of the GEF's PES projects.

Probably as a result of GEF involvement, the ultimate goal of almost all the PES projects reviewed was biodiversity conservation. In contrast with private PES schemes that tend to focus on a single environmental service, projects in the GEF portfolio typically involve a bundle of environmental services. Watershed management for water quantity and quality, carbon sequestration and landscape conservation for tourism are the ecosystem services that are expected to trigger payments (rather than biodiversity conservation).

AN OVERVIEW OF KEY DESIGN ELEMENTS

Who Leads the Design?

According to GEF bylaws, recipient country governments should be the ones to prioritize needs, identify preferred interventions and lead engagement with prospective partners and a GEF implementing agency to design projects that will vie for GEF grants. The portfolio shows evidence of a strong leadership by governments, particularly in Latin America, a region that has been at the forefront of PES experimentation. For example, the largest projects in the portfolio in Costa Rica and Mexico were initiated by their governments, well before the GEF came on board.

The GEF grants funds, but projects are developed and implemented by the GEF's implementing agencies: the World Bank, the United Nations Development Programme (UNDP) and the United Nations Environment Programme (UNEP).[4] This arrangement ensures the pooling of funds and experience, though the GEF has a significant influence in overall portfolio design. The GEF Council decides the funds' allocation strategy (where and on what the GEF money will be spent), the GEF Board approves each grant (and many proposals are turned down) and the GEF Secretariat reviews each project proposal (and usually asks for many changes to the project design).

The World Bank takes the lead on country-level grants associated with investment projects, the UNDP leads on multinational projects and the UNEP leads on science-focused projects. Since GEF members have always directed the majority of GEF budget to country-level investment projects, the World Bank is guaranteed a leading role. This is true both for the overall GEF portfolio and for the GEF PES portfolio, where the World Bank implements two-thirds of the projects and 80 per cent of the budget (FAO 2007).

In practice, based on the authors' experience with these institutions, it is likely that projects managed by the World Bank were designed to a large extent by World Bank staff in consultation with country government officials, whereas projects managed by UNDP and UNEP were designed either by independent consultants in collaboration with the institutions that would receive the GEF grant, or by the latter themselves.

A grant, and the participation of implementing agencies, makes a difference to the design of PES schemes. Where an existing PES was funded by GEF – as in Costa Rica and Mexico – the GEF participation encouraged better social and environmental targeting, the diversification of buyers and better monitoring of outcomes. However, it is uncertain as to whether implementing agencies affect project design, because implementing agencies – with the exception of the World Bank – have only one or two PES projects. The World Bank has the resources, motivation and technical expertise to design detailed PES-focused projects with significant consideration given to balancing environmental and social goals, which is the case in the four World Bank PES projects among the six stand-alone projects identified in Table 2.1. Other GEF implementing agencies lack the in-house PES expertise of the World Bank, usually deferring the project design to on-the-ground proponents while retaining an evaluation role; thus the project's final design can be quite sophisticated. That is the case for the projects in Uganda (implemented by UNEP) and the global one (implemented by UNDP); both championed and designed by the non-government organization (NGO) that will undertake them.

More generally, all GEF projects are required to enlist the collaboration of local organizations (including civil society organizations, NGOs, local authorities, trade associations and business) both in initial consultations regarding PES design and during project implementation. The limited evidence available suggests that this has been done to a significant degree, with demands from the public leading, for example, to the redesign of PES programmes.

The Buyers of Ecosystem Services

Currently, the largest buyers of ecosystem services in the GEF portfolio are, in descending order: national governments, either through the earmarking of specific revenues (for example, a fuel tax or a water fee) or from budgetary allocations; the GEF and other international donors; and private buyers, in a distant third place. In most cases, the payments go first to an agency or an environment fund which then pays environmental service providers.

With the exception of the Costa Rican and Mexican national, government-led PES schemes, the demand for ecosystem services is expected to arise, in the long run, from private national and international buyers. Such buyers would be interested in water- and carbon-related ecosystem services and, to a lesser degree, in landscape conservation for eco-tourism, and biodiversity conservation. Table 2.2 summarizes the available information on the current and expected ecosystem service buyers.

In 16 projects, PES will be one of several income streams. For example, in the Dominican Republic's Yegua Watershed System Project, the

creation of a Watershed Environmental Fund is envisioned to pay farmers
that adopt sustainable land-use management practices and collect funds
from several sources, including environmental service payments (irrigation users, potable water users and hydroelectric plants), debt-for-nature
swaps, a risks guarantee fund and traditional donations and fund-raising.

Most PES projects indicate that GEF funds will be used to pay for institutional development and capacity-building, for design and start-up costs,
and for that part of the conservation effort that is of global significance. In
several of the small-scale projects, all expenses, including the actual payments for ecosystem services are made out of the GEF and other donors'
grants, a situation that raises concerns about sustainability beyond the
projects' end.

The design of some projects is very vague about who would actually pay
for the environmental service, stating that buyers will be identified some
time in the future, or putting forward a wish-list of funding sources. This
occurs primarily, but not only, in projects where PES is just one of several
financial alternatives.

The Sellers of Ecosystem Services and the Projects' Social Targeting

Current and would-be providers of ecosystem services are primarily small
farmers; medium and large landowners come in a distant second place,
and a few projects have protected area agencies as a major seller of ecosystem services (see Table 2.2). In all except one project, payments are made
in cash.

With respect to social and environmental targeting (see Table 2.3)
three distinctive groups were found in the GEF PES portfolio. The first
group included projects that target both the rural poor and areas with
high environmental service potential. To a great extent, national PES
projects or demonstration projects (Uganda, Costa Rica, Mexico and the
regional Colombia, Costa Rica and Nicaragua projects) can choose where
to work and so it is easier to identify areas that have both high environmental services potential and high potential for the rural poor to become
sellers of ecosystem services. Still, the results have been mixed on both the
environmental and social fronts.

The second group included projects whose primary focus is on environmental protection, and do not have a social component – as in the case
of South Africa's Cape Action for People and the Environment and the
National Grassland projects – where the relevant natural resources are
owned by medium and large landowners, so there is little opportunity
there for the rural poor to become providers of ecosystem services.

The third group comprises rural development projects which typically

Table 2.2 *Buyers and sellers of ecosystem services*

Region	Buyers					Sellers			
	Governments	Water users	Carbon sequestration buyers	GEF	Others/not defined	Small Farmers	Medium and large landowners	Protected Area agencies	Others/not defined
Africa	1	4	–	1	4	3	2	1	1
Latin America	7	11	7	5	7	15	3	3	–
Global	–	1	1	–	1	1	1	–	1
Total	8	16	8	6	12	19	6	4	2

Source: GEF (2008) projects database.

33

Table 2.3   Social targeting

Region	Targeting				Social groups identified as targets		
	Both environmental and social target	Mostly environmental target	Mostly social target	Rural poor	Rural communities	Indigenous population	Women
Africa	2	2	1	2	3	1	-
Latin America	8	4	4	16	4	6	3
Global	1			1	1	1	1
Total	11	6	5	19	8	8	4

Source: GEF (2008) projects database.

have a high level of social targeting but a lower emphasis on the environmental outcome – for example, Kenya's project and Brazil's Sustainable Land Management in the Semi-arid Sertao project. One should not conclude from this that rural development projects, or those that have a strong social focus, will inevitably have a poorly designed PES component. Rather, in these projects PES is at an exploratory stage and will likely play a small role, if any.

Beyond the rural poor, many projects identify more specific social targets, including rural communities, indigenous communities and, to a lesser degree, women (see Table 2.3). This targeting is achieved in some cases by designing the whole PES project to focus on a single group – for instance, the Mexico PES buys ecosystem services only from rural communities – or, more commonly, by giving additional support to participants from the targeted groups – for example, women heads of households in rural Costa Rica.

Land Use or Management Changes to Deliver Ecosystem Services

Monitoring the actual flow of ecosystem services can be difficult, either due to the lack of a clear baseline, or due to time lags and random fluctuations in natural cycles. Therefore, farmers are usually paid to implement a land-use change that is positively related to the flow of the desired environmental service. Such land-use changes include protection of natural forests, reforestation, silvopastoral activities, agroforestry, other sustainable agriculture practices, protected area management and buffer zone management.

THE COSTA RICA PES EXPERIENCE

PES Design

The PES scheme of Costa Rica encompasses the entire country, and was originally designed to pay for reduced deforestation and increased reforestation, with the goals of increasing the supply of carbon sequestration, the maintenance of forest hydrological services, biodiversity conservation and the preservation of scenic beauty. The programme started in 1997 and came into being under national legislation enacted in 1996. It was supplemented by funding from the GEF in 2000. The GEF helped to support the biodiversity aspect of the programme and increased the focus on gender equity issues and participation by indigenous communities (World Bank 2008).

The creation of the Costa Rica PES scheme was preceded by a period of national debate in which a number of interlinked environmental problems were discussed. Deforestation had been recognized as a serious problem in Costa Rica since at least the 1970s. By that time, the country had one of the highest rates of deforestation in the world, and this high rate continued through the 1980s as the forests, which had covered 50 per cent of the country in 1950, were reduced by half by 1995 (World Bank 2007). Initial efforts to address this situation were in the form of the government's forest certification programmes that first promoted timber plantations and were later modified to support forest conservation (Chomitz et al. 1998). The *Pago por Servicios Ambientales* (PSA, Spanish for PES) programme was built on the institutional foundation of the certification programmes, but changed the focus from timber industry support to PES. The change in scope reflected an effort to value the goods and services that forests provide beyond the harvestable timber (Pagiola 2008).

The legislation governing Costa Rica's PSA programme, *Forest Law No. 7575*, created the National Forest Finance Fund (*Fondo Nacional de Financiamiento Forestal*, FONAFIFO) which was empowered to buy environmental services from private landowners. FONAFIFO is an independent hybrid legal entity whose governance involves both the public and private sectors, which acts as an intermediary, contracting with ecosystem services providers (namely private forest land owners) and with fiduciary responsibility for funds collected from the government and other users of ecosystem services (de Man 2004; FONAFIFO 2008). The primary source of funding for the PES scheme is a tax on gasoline. There is also support from the Ministry of the Environment and Energy (MINAE) and substantial funds from international donors and multilateral lending agencies, including the GEF, and small but growing payments from water users, both on a voluntary basis by some large water users and through a mandatory tariff recently added to the water bill.

The price paid for ecosystem services is determined by FONAFIFO with input from many social sectors and stakeholders. Would-be sellers must prepare a sustainable forest management plan in conjunction with a certified forester, a process that is sometimes facilitated by an NGO (Pagiola 2008). The final sellers are then selected from a large pool of applicants who have land in priority areas (IIED 2007).

Payments are made for forest conservation, reforestation, agroforestry and natural regeneration activities, in order to improve the delivery of carbon sequestration, water quantity and quality, biodiversity and scenic beauty. Landowners are the recipients of the payments, which vary by the type of forest activity in which they engage. Payments for forest conservation were US$64/ha/year in 2006, based on 5-year contracts that could be renewed once. Payments for reforestation were US$816/ha in 2006, over a 10-year contract, with most of the payment front-loaded in the first year. Payments for natural reforestation of degraded land or pastures were US$41/ha/year over 5 years in 2006, with per tree payments of US$1.30 for agroforestry contracts (FONAFIFO 2008; FAO 2008).

Sellers of ecosystem services have an obligation to demonstrate that their forestry land-use plan is sustainable and they commit to the plan's land-use provisions for a period of 5 to 10 years. This commitment is transferred with the land if it is sold (de Man 2004). The programme includes significant sanctions for non-compliance, and compliance inspection and monitoring is the responsibility of the *Sistema Nacional de Areas de Conservacion* and MINAE (Sanchez-Azofeifa et al. 2007) with cooperation from NGOs (World Bank 2000). The penalty for non-compliance is forfeiture of future payments, while certified foresters who incorrectly certify compliance put their licences at risk (Pagiola 2008).

Transaction costs can be significant, particularly for small landowners (Pagiola 2008). Fees for the development of sustainable management and subsequent mandatory management engaged by the landowner either directly or often through an intermediary are estimated to range from 12–18 per cent of payments (Grieg-Gran et al. 2005, Pagiola 2008). Intermediaries include the Foundation for the Development of the Central Volcanic Mountain Range (FUNDECOR), an NGO, and the municipally owned utility *Empresa de Servicios Públicos de Heredia* (Miranda et al. 2003). In the Virilla watershed, 80 per cent of those surveyed used an intermediary to access the programme, possibly because of the significant complexity in terms of documentation and time commitments (Grieg-Gran et al. 2005), although by some estimates direct access by the landowner would in some cases be less costly (Miranda et al. 2003). Another significant transaction cost is the requirement that land be kept out of production from the time the application for participation is submitted until it is accepted, which may be onerous for some smallholders leading them to forgo participation.

The Demand

The sources of the programme's financial support have implications both for the scale of the programme's impact and the levels of participation, both of which are dependent on confidence in the fiduciary soundness of the programme.

Water services

Initially water service payments were voluntary, with individual water users paying FONAFIFO, who then paid upstream landowners for maintaining or improving watershed protection in forests. The first buyers to enter into this type of arrangement were hydroelectric power producers in the late 1990s. As the programme was streamlined, the number and diversity of buyers grew to include bottlers, a municipal water supply company and several agribusinesses (Pagiola 2008; FONAFIFO 2008). With the revision of Costa Rica's water tariff in 2005, payments for watershed conservation are a mandatory part of the water tariff, a conservation fee that is specifically set aside for water conservation, with 25 per cent of the fee going to the PSA programme. While the general nature of this national tax tends to weaken the financial link between service providers and users, the programme works to maintain the link by ensuring that tariffs collected in a given watershed are used in that same watershed, and by giving large users the option of opting out of all or part of the mandatory tariffs and entering into voluntary direct payment agreements with FONAFIFO (Pagiola 2008).

Carbon payments

In the Costa Rican PES scheme, payment for carbon comes from FONAFIFO and is based on funding from a tax on fossil fuel and the sale of offsets in the international carbon market. The connection between the fuel tax and actual carbon sequestration is somewhat diffuse, in that the tax probably falls short of what would be needed to fully offset Costa Rica's fossil fuel carbon emissions, and revenues are included in the general PSA fund.

Finance from the sale of carbon emission offsets is more straightforward. For example, the Emission Reduction Payment Agreement concluded with the World Bank's BioCarbon Fund, specifies that the payments are for quantified carbon sequestration activities by a cooperative of farmers in a particular location (World Bank Carbon Fund Unit 2008). FONAFIFO first entered the voluntary carbon market with Certifiable Tradable Offsets (CTOs) which, although developed by FONAFIFO itself, were externally certified (Chomitz et al. 1998). The CTOs were initially successful, with sales to Norwegian power producers (Thomas 1998). However, rules associated with the creation of the Clean Development Mechanism (CDM) excluded forest conservation as a creditable activity, so the international demand for biocarbon did not grow as fast as Costa Rica had expected. This limited the scope of projects such as the one in the BioCarbon Fund portfolio and the range of international funding options has, thus far, remained limited.

Biodiversity payments
At the beginning of the 1990s, Costa Rica bet strongly on the development of an international market for bioprospecting and put in place several innovative arrangements to attract foreign buyers. Unfortunately, the market did not develop and the biodiversity component of Costa Rica's PSA system is currently supported by GEF grants (FAO 2007). Although the GEF funds many activities associated with specific biodiversity conservation in protected areas, payments are not directed to specific activities at specific locations as is the case with the external funding for carbon sequestration. Though diffuse, the GEF payment is a recognition that the PSA makes a significant contribution to the conservation of biodiversity, a global environmental good. The GEF thus acts in part as the intermediary between the international community (the buyers) and the PSA participants (the sellers), but the linkage is not direct, GEF funds are limited, and a more sustainable funding source for biodiversity is yet to be found.

Provision of scenic beauty
Thus far, Costa Rica's extensive ecotourism industry has not stepped forward as buyers of the scenic beauty ecosystem service in the way that some hydroelectricity producers have done with water services (Pagiola 2008). This may be due to the tourism and ecotourism industry profiting as free riders, but research suggests that, as it is currently designed, there are few ecotourism benefits from the PES programme (Ross et al. forthcoming). This conclusion was based on the results of an economic model where the tourism value of a hectare of forest was estimated at US$16/ha/year.

The Supply and the Social Impact

The supply of ecosystem services comes from landowners that apply to the programme on an individual and voluntary basis. Thus far, most participants tend to be well educated, understand the PSA system, live in urban areas and have existing incomes at the high end of the economic spectrum, much of it from off-farm sources. Participants also tend to have higher farm incomes and higher debt levels, and their land tends to have more degraded soils than the average (Zbinden and Lee 2005).

With respect to the programme's social targeting, the initial PSA design hindered poor farmers' participation. For example, parcels acquired through land reform were excluded from the programme, and participants were, in some cases, excluded from other government-supported programmes such as housing subsidies and faced restrictions in the access to other ones, such as credits from the National Banking System (Grieg-Gran et al. 2005). With the participation of the GEF project, there have been systematic efforts to make the PSA scheme more accessible to the poor, to indigenous communities and female-headed households. These efforts included relaxing the requirement for clear title and reducing the complexity of the application process in order to reduce transaction costs. Consequently Hartshorn et al. (2005) report significant improvements in the participation of these three groups.

As for the programme's impact on incomes, since payments are on a per hectare basis, large landowners receive larger total payments, but the actual livelihood impact depends on the proportional contribution to family income. De Man (2004) found that PSA payments could increase household incomes by as much as 50 per cent for poorer farmers, although the perception among these farmers was that the actual impact was much less. The same study found that participants were happy to have the PSA programme as a new source of income and felt that it created value in forests that had not existed before. Another study found that participants viewed PSA payments as making a significant improvement in income diversification and stabilization as well as off-setting the costs of project start-up (Grieg-Gran et al. 2005).

Impacts on the livelihood of other groups – for example, buyers of ecosystem services and rural dwellers whose access to resources or job opportunities may be affected by the PSA programme – have not been assessed as yet. Ross et al. (forthcoming) modelled the economy-wide effects of the PSA programme in several possible future scenarios and found the effects to be very small, a result that was independent of the quantity of land or funding levels. However, it is important to note that the specific effects of the PSA programme at the regional, local or familial level could be significant, even if the effects were small at the national level. Where forest cover is high and incomes are low, the local economic contribution could be proportionally high. For example, PES increases in household income in the Aranjuez watershed were estimated to be as high as 50 per cent for

low-income participants, while in the Virilla watershed the PES contribution to household income has been estimated to range from 4 to 34 per
cent of household budgets, depending largely on size of the landholding,
with higher values for larger landholdings (Miranda et al. 2003).

Environmental Impacts

The environmental impacts of the PSA programme have been the subject
of substantial research efforts, but controversy lingers. Tattenbach et al.
(forthcoming) estimated that the PSA programme had reduced deforestation between 1999 and 2005, such that forest cover was 10 per cent higher
than it would have been in the absence of the programme. Hartshorn et
al. (2005) also found success in terms of reduced deforestation and social
impacts in the evaluation the first of the two GEF projects, but were concerned that the programme did not efficiently target the most vulnerable
areas. Similarly, Sierra and Russman (2006) found that inadequate targeting of habitats at greatest risk had significantly reduced the environmental
benefits of the programme and called for revised payment criteria.

Sánchez-Azofeifa et al. (2007) argued that Costa Rica had made significant progress in reducing its rate of deforestation prior to the inception
of the PSA programme and that, after its inception, further reductions in
deforestation were small but significant. They also examined the relationship between proximity to the deforestation front and PSA payments and
found that lands near the deforestation front were only slightly more likely
to receive payments. Thus the payments were not focused on areas where
deforestation risk may be higher. The positive impact on biodiversity was
somewhat higher, with somewhat greater participation in the PSA for
lands within biodiversity conservation corridors as compared with areas
outside these corridors.

Costa Rica: the Broader Context

The Costa Rican PES experience represents an early and arguably successful effort to implement a PES scheme that combines environmental
and social benefits. The programme was initiated in response to a socially
perceived need and enhanced by the support of the GEF. However, efforts
to find private buyers for ecosystem services have met with only limited
success and the programme remains largely dependent on tax revenues. The
programme has had some success in combining environmental and development benefits, but in each case the targeting of areas of greatest need could
be improved. The scale of the programme in the context of the national
economy is small, but impacts could be significant at a household level.

MEXICO PSA-H EXPERIENCE

The PSA-H Design

Mexico's Payments for Forest's Hydrological Environmental Services Programme (*Program de Pagos por Servicios Ambientales Hidrológicos de los Bosques*, PSA-H) is a PES scheme developed by the government of Mexico, with a focus on increasing fresh water availability, particularly groundwater. The scheme also aims to have positive socioeconomic impacts on marginalized communities by focusing on *ejidos*, rural communities on legally recognized communal lands that are an outgrowth of the Mexican revolution. The impetus for the programme was the recognition that water quantity and quality were at risk due to deforestation in upstream locations, particularly where commercial forestry was less lucrative and not competitive with livestock and agricultural production (Bezaury-Creel and Gutierrez 2007) as this may have been a deforestation driver.

The PSA-H was initiated in 2003 based on extensive studies by the Instituto Nacional de Ecologia, the Universidad Iberoamericana, the Centro de Estudios y Docencia Economica and the University of California at Berkeley. Scientific studies provided the basis for the legislation; however, the political process whereby the legislation was approved was complex and lengthy (Muñoz-Piña et al. 2008). The outcome was the establishment of the PSA-H, which is the responsibility of the National Forestry Commission (CONAFOR). CONAFOR designs the programme in consultation with stakeholders, performs monitoring, oversight and evaluation functions, and holds technical and financial responsibility for the programme.

In Mexico, water is (legally) recognized as a public good and the government acts as the intermediary between the landowners that may contribute to the provision of water quantity and quality, and water users, including private companies, municipalities and households. The PSA-H programme is national, but it is the stated intention of the national government to increase participation by municipal governments, with the expectation that they will gradually take on some of the programme's financial responsibilities.

Geographic and ecological criteria are the primary means by which ecosystem services sellers are selected under the PSA-H system. Priority is given to the recharge areas of overexploited aquifers, which should also be in watersheds experiencing great water scarcity and where the risk of flooding is great. Cloud forests – which are viewed as sources of water capture and recharge – are given greater value than other forest types. In order to be eligible, forests also need to be near a town or settlement of at least 5000 people. After stakeholder consultation, protected areas and designated priority mountains were added to the list of eligible areas. CONAFOR is responsible for selecting participants and disbursing payments. Sellers, once selected, have the responsibility to halt deforestation on private and communally held land (Muñoz-Piña et al. 2008).

In 2004, the Mexican Government added a smaller parallel programme focusing on carbon sequestration and biodiversity conservation (PSA-CABSA, see Chapter 3) and, in 2007, with support of the GEF project, a Biodiversity Fund for the protection of biodiversity of global significance was also added to the set of Mexico's national PES schemes.

The Demand

In order to have a full understanding of the PSA-H programme, the way it functions and, ultimately, its potential sustainability, it is necessary to understand its client demand and its sources of funding.

In addition to the GEF grant, funds come from the national water tax, approximately 2.5 per cent of which is apportioned for the PSA-H programme. In recent years, revenues from this source have amounted to between US$18 and US$28 million per year (Muñoz-Piña et al. 2008; Bezaury-Creel and Gutiérrez 2007).

The price paid to landowners to conserve forests was originally set by CONAFOR, based on a study of the opportunity cost of alternative agricultural and livestock land uses. During subsequent stakeholder consultation, the original prices were deemed by stakeholders to be too low and were raised – despite the fact that, within the fixed budget, this would reduce the area covered (Muñoz-Piña et al. 2008).

Agreements are for 5 years, with annual cash payments. In 2005, the payments were approximately US$36.40/ha/year for cloud forests and US$27.30/ha/year for all other forests. The sanction for non-compliance (that is, deforestation on contracted land) is cancellation of current and future PSA-H payments. Transaction costs are, for the most part, subsumed within CONAFOR management, and technical support is provided as part of the national effort at forest management. In the three years 2003–2006, contracts were signed over 600000 hectares (Muñoz-Piña et al. 2008; Martinez 2007, Bezaury-Creel and Gutierrez 2007).

Carbon sequestration services under PSA-CABSA are currently paid by the Mexican Government with the expectation that international demand for biocarbon will increase in the near future. During 2004–2006, the Mexican Government invested approximately US$16 million in the programme both in direct payments and project development. In 2007, payments per hectare went from a one-time payment of US$368/ha for carbon sequestration, to US$30/ha/year for agroforestry projects (Bezaury-Creel and Gutierrez 2007; Martinez 2007). The PSA-CABSA programme also includes a small component of payments for biodiversity protection (Bezaury-Creel and Gutierrez 2007; World Bank 2006). Global biodiversity payments came from a Biodiversity Fund that was established in 2007, partially capitalized with the GEF grant. In its first round, payments have been set at approximately US$18/ha/year.

Social Impacts

Both PSA-H and PSA-CABSA were designed with strong social targeting. As many *ejidos* are highly marginal economically, and as marginality can be considered a proxy for poverty, the Mexican Government decided that directing payments to *ejidos* located in priority conservation areas would guarantee strong social targeting. Originally 100 per cent of payments were made to *ejidos* but this was subsequently modified so that 90 per cent of the resources go to ejidos and 10 per cent go to private landowners (Alix-Garcia et al. 2005).

   Still, in the initial year of the PSA-H programme, a tight implementation deadline resulted in a participant selection process that was largely ad hoc. There was also a misinterpretation of programme rules that led to the use of forest cover rather than forest density as a selection criterion (forest cover is the spatial extent of all areas classified as forest regardless of density, whereas forest density is the actual density of trees). This tended to favour the inclusion of larger properties with lower population densities, which were at lower risk for deforestation as they had larger areas of forest cover by virtue of their size. In the following years, a more regularized and transparent selection system was implemented, through the formation of a technical committee and based on the allocation of points (Alix-Garcia et al. 2005).

   Communities receiving payments used the money in a variety of ways. Some divided it equally among community members, others used part of the money for conservation activities, infrastructure (road repair, radio communications) or other community activities such as school improvement (Alex-Garcia et al. 2005). Some experts (Pagiola et al. 2005) have raised the concern that community-level payments may allow for inequities among the community members, for example, through unfair distribution of incomes or differential access to communal resources. Recent studies conclude that the PSA-H programme was very effective at targeting payments to the rural poor (Muñoz-Piña et al. 2008; Alix-Garcia et al. 2008). Within communities, there were also strong tendencies toward equitable distribution of benefits supported by civil society, often through the creation of common goods, using payments such as school improvement or micro-finance (Alix-Garcia et al. 2008). Thus, in practice, cultural norms of fairness and equitable distribution seem to help ensure against inequities within the communities. The livelihood impacts on non-participants have not been quantified.

Environmental Impacts

Mexican PES programmes seem to do better socially than environmentally, and there may be a trade-off between these two goals. In some instances, environmental goals have not been optimally achieved. Several studies have found that areas at very high or high risk of deforestation are currently under-represented in the programme, while areas of low or very low risk of deforestation are over-represented. Similarly, overexploited aquifers were under-represented in the payment areas of the current PSA-H programme, while the majority of payments go to areas where aquifers are in equilibrium or had a margin for expanded use (Muñoz-Piña et al. 2008).

Mexico: the Broader Context

The PSA-H programme, although national, has had limited impact. However, the potential for scaling up is now in place as there are now several years of institutional experience on which to build. The PSA-H programme provides an important example in the broader PES discussion as it presents an example of direct payments to communities that are collectively controlled, providing the potential for local level empowerment of existing community-based institutions in addition to the other benefits that a PES scheme can provide. The PSA-H programme also demonstrates the potential power of communities in influencing the form of a PES scheme through early and direct participation in scheme development. While the PSA-H scheme is publicly funded, the explicit inclusion of carbon as a value within the set of PES schemes could provide a catalyst for future participation in international carbon markets.

DISCUSSION

Project Scale and Project Impact

On the basis of evidence from the previous case studies, it would appear that PES projects in the GEF portfolio are not likely to become a major environmental tool and a significant source of income for rural house-holds. In Costa Rica, after 10 years of paying landowners to conserve natural forests, the area under contract is significant – some 15 per cent of the country's natural forests – yet annual payments, of US$12 million a year, represent less than 0.6 per cent of the country's rural income. In Mexico, which has spent US$20–30 million per year in its PES programme, the area under contract by late 2006 was less than 1 per cent of the country's natural forests and the annual budget for PES was less than 1 per cent of the rural subsidies distributed annually by the Federal Secretary of Agriculture. Nevertheless, even if they are small at a national scale, PES projects may be an important source of income at the local scale.

Sustainability and Scaling Up

In theory, PES arrangements could be sustainable, provided ecosystem services are correctly valued and the buyers and sellers are satisfied with the arrangement. In practice, PES schemes have been in existence for such a short time that their true sustainability remains unknown. The short project cycle is of concern, though the issue is far from being resolved – most GEF PES projects go no further than funding PES design and start-up costs, and hope that by the close of the project they leave behind a functioning PES scheme with sellers and buyers in place.

Although national projects that are financed largely through earmark-ing of taxes (the fuel tax in Costa Rica) or tariffs (water tariffs in Mexico) may enjoy a strong financial floor, limited funding, poor environmental and social targeting, as well as capture by political agendas are matters of concern. Biodiversity conservation usually fails to find private buyers, thus most GEF PES projects propose to capitalize a biodiversity fund (for example, South Africa Cape biodiversity, Mexico Environmental Services Project). Still other projects (for instance, the Colombia, Costa Rica and Nicaragua silvopastoral project, the Brazil Sustainable Land Management in the Semi-arid Sertao) were designed on the assumption that PES payments were required only for the duration of the project – to pay farmers for the upfront costs of adopting sustainable agricultural or agroforestry practices – and that, once adopted, would prove profitable, so that farmers would not require or demand further payments to keep providing the ecosystem services in question.

Scaling Up

The GEF's role in the PES arena looks small, but is nevertheless impor-tant in several ways – acting as the motivator for other institutions to participate, funding institutional development and capacity-building, and promoting new design ideas and approaches – which may also be relevant for REDD payments in the future. However, in terms of environmental and social impact, the GEF PES portfolio is small and there is no consensus on what, if any, lessons for scaling up can be derived from it, although the institutional learning engendered in the national PES systems suggests that scaling up might be possible.

Narrow definitions of PES would restrict the PES label to private market, voluntary and conditional transactions; and only for ecosystem services that perform as economic externalities (see Wunder et al. 2008). Besides carbon sequestration, practitioners that endorse this narrow approach to PES see potential only in replicating private markets for watershed-related ecosystem services. However, even in the most optimistic scenario, private markets for watershed ecosystem services tend to be small and local, and as every watershed is different, they tend to have high start-up and transaction costs and therefore offer limited lessons to PES developments elsewhere. It is the authors' view that narrow approaches to PES will miss the majority of ecosystem services (for example, African community wildlife conservancies or markets for environmentally friendly products) and will limit PES to the marginal role that it currently has (see Gutman 2007).

Livelihood Impacts

PES programmes in the GEF portfolio can and are delivering livelihood benefits to poor rural communities, but these benefits are modest, considering that a rural poor household owns, at best, a few hectares of land and payments for forest conservation have been relatively low (between US$27/ha/year in Mexico and US$64/ha/year in Costa Rica). Even within a PES programme there can be very different livelihood impacts from one locality to another: for instance, in the Aranjuez watershed in Costa Rica, the PES contribution to participant poor households was estimated to be up to 50 per cent of household income (de Mann 2004), while in the Virilla watershed, participation in the PES programme boosted smallholders' income by as little as 4 per cent (Miranda et al. 2003).

The Costa Rican and Mexican PES programmes illustrate some of the efficiency and equity challenges that national programmes face, stemming in part from national and local realities that frame PES design options, and in part from the inherent complexity of these schemes and the potential for unintended consequences. There are many technical options to improve social and environmental targeting (see, for example, Alix-Garcia et al. 2008) and explicit analysis of potential trade-offs needs to be incorporated into the rule-making process to ensure that they are equitably addressed. The relative capacity of different environmental service schemes to provide support for social and environmental benefits is an area in need of active research as the limited examples do not yet provide an explicit answer to this question.

LESSONS FOR THE DESIGN OF PRO-POOR REDD PROJECTS

The GEF PES portfolio (especially the Costa Rican and Mexican experiences) offer several valuable lessons about whether, and how, national REDD schemes will be able to provide benefits to the poor. The GEF programmes show that, with the necessary logistical and organizational infrastructure in place, countries such as Costa Rica and Mexico have effected payments to smallholders either directly or through payments to communities. Important preconditions include government capacity to produce enabling legislation; the political will to implement the legislation; effective land-use survey methods; and contact with smallholders either directly or at the community level.

The GEF portfolio also highlights the kinds of institutional capacity-building that would be necessary, including a functioning and consistent legal system with secure land tenure, capacity on the part of government agencies, ministries and dedicated entities to handle and disburse funds, and the technical capacity to accurately and consistently quantify land-scape information. Such capacity-building will require understanding and recognition within the country's political structure of the potential of PES as a conservation and development strategy.

GEF experience demonstrates that national programmes can meet demands from (international) buyers interested in purchasing services from a particular location or targeting specific social groups; targeting can be achieved with credible monitoring and reporting standards. GEF experience also shows that focusing on marginal areas where poverty is high can improve social targeting.

GEF PES experience can also inform questions of how much to pay, in what form and to whom. The preferred approach in the GEF PES portfolio is to pay for the forestland's opportunity costs, though it is difficult to estimate these opportunity costs and devise payment schemes that approximate these costs well. When the flow of ecosystem services is linked to the adoption of sustainable agriculture or agroforestry practices, paying for input costs is the preferred approach. Cash payments to individual farmers are the most common practice, but PSA-H in Mexico shows that negotiating with and paying rural community institutions is a reasonable option, provided that the rural institutions have the legal right and social standing to represent the local land users. Using this model could also reduce programme transaction costs, in some cases, by reducing the number of sellers to track as well as having the potential to further empower communities within the programme context.

It has sometimes been argued that only by engaging with large landowners or large organizations of landowners could PES schemes maintain reasonable transaction costs. Although information about transaction costs is limited, the fact that transactions continue and that sellers are satisfied with their payments demonstrates that making payments to multiple participants, many of them smallholders, can be achieved with transaction costs that do not exceed 20 per cent of PES payments. Thus national schemes can be effective in providing participation access to smallholders. This is particularly important as many REDD proposals are targeted at the national level.

Most poor farmers and rural communities will have too little land to make a living from REDD payments alone, so they are likely to have to allocate their limited land and manpower resources among several income-generating activities. The GEF portfolio of mixed PES projects may provide important lessons, where PES is one of several funding mechanisms for sustainable land-use management. These projects try to combine PES with traditional sources of rural income (for example, increased productivity and increased market access) and PES-related sources (for instance, markets for certified organic or environment-friendly products; eco-tourism, etc.).

The timescale for, and conditionality of, payments will need to be addressed by both PES and REDD. The idea of an open-ended conditional commitment has made PES difficult to sell, and it is interesting to note that several projects in the GEF PES portfolio front-load payments or offer renewable contracts (of limited duration). The assumption in both cases is not that the ecosystem service will end with the end of the payments, but rather that a limited payment period is enough to secure the long-term delivery of the environmental service in question. Managing ongoing carbon sequestration in REDD poses similar challenges in designing payment schedules and conditions.

From a rural poverty alleviation perspective, it may be worth considering REDD as part of a transition to more inclusive carbon markets that fully credit the whole range of agriculture, forest and other land use (AFOLU) activities that sequester carbon. The world REDD market, as originally envisioned, will concentrate on a small number of tropical countries that currently have large forests and high deforestation rates (see Ebeling 2006). Hundreds of millions of peasants and small farmers in Africa, Asia and Latin America living elsewhere would have little to offer in terms of REDD, but could still contribute significantly to climate change mitigation through AFOLU-related carbon markets. Further, to make REDD fully effective, it will probably be necessary to consider the forest in a broader landscape that includes the adjacent mixed use and agricultural landscapes which provide a broad range of livelihood possibilities that can supplement REDD finance and provide an incentive to maintain the integrated landscape sustainably.

NOTES

1. This chapter draws heavily on previous work by P. Gutman and S. Davidson (FAO 2007).
2. The real figures may be substantially lower, given that projects where PES is only a small component of a larger project account for two-thirds of the GEF PES portfolio (see Table 2.2).
3. Latin America leads in number of projects in all surveys of PES experiences in developing countries (for example, Landell-Mills and Porras 2002), but not by as a much as it does in the GEF PES portfolio.
4. In 1999 the GEF enlarged its partners' pool, designating as executing agencies the four regional development banks plus FAO, IFAD and UNIDO. Thus far, they account for a small number of GEF projects.

REFERENCES

Alix-Garcia, J., A. de Janvry, E. Sadoulet and J.M. Torres (2005), *An Assessment of Mexico's Payment for Environmental Services Programme*, Rome: FAO.
Alix-Garcia, J., A. de Janvry, E. Sadoulet and J.M. Torres (2008), The role of deforestation risk and calibrated compensation in designing payments for environmental services. *Environment and Development Economics*, **13**, 375–94.
Bezaury-Creel, J. and L. Gutierrez (2007), El papel de los servicios ambientales para evitar la deforestación en Mexico, in *Servicios de Ecosistemas en América Latina. Lecciones Aprendidas en Agua, Bosque y Ecoturismo*, Cartagena de Indias, Colombia: The Nature Conservancy, USAID, A.C. Walker Foundation, pp. 14–24.
Chomitz, K., E. Brenes and L. Constantino (1998), *Financing Environmental Services: the Costa Rican Experience and its Implications*, Washington DC: Development Research Group, World Bank.
de Man, M. (2004), Local impacts and effectiveness of payments for environmental services in Costa Rica: the case of payments for forest hydrological services in Costa Rica's Aranjuez watershed, MA Thesis, Utrecht University, NWS0-I-2004-37.
Ebeling, J. (2006), Tropical deforestation and climate change. Towards an international mitigation strategy, MSc Dissertation, Oxford University, UK.

FAO (2008), Payments for Environmental Services from agricultural landscapes: targeting providers and accomplishing additionality, available at http://www.fao.org/ES/ESA/pesal/PESdesign6.html (accessed 1 December, 2009).

FAO (2007), The Global Environmental Facility and Payments for Ecosystem Services. A review of current initiatives and recommendations for future PES support by GEF and FAO programmes, report commissioned by FAO for Payments for Ecosystem Services from Agricultural Landscapes- PSAAL project, PSAAL Papers Series No.1 Rome.

FONAFIFO, (2008), The website of the Costa Rica PES agency, available at www.fonafifo.com (accessed 10 October 2008).

GEF (2008), GEF Project database, available at http://gefonline.org/home.cfm (accessed 21 September 2008).

Grieg-Gran, M., I. Porras and S. Wunder (2005), How can market mechanisms for forest environmental services help the poor? Preliminary lessons from Latin America, *World Development*, **33** (9), 1511–27.

Gutman, P. (2007), Ecosystem services: foundations for a new rural-urban compact, *Ecological Economics*, **62**, 383–7.

Hartshorn, G., P. Ferraro and B. Spergel (2005), *Evaluation of the World Bank-GEF Ecomarkets Project in Costa Rica*, Raleigh, NC: North Carolina State University, USA.

International Institute for Environment and Development (IIED) (2007), Watershed Markets, available at www.watershedmarkets.org/casestudies/Costa_Rica_National_PSA_eng.html (accessed 1 October 2008).

Landell-Mills, N. and I. Porras (2002), *Silver Bullet or Fool's Gold? A Global Review of Markets for Forest Environmental Services and their Impacts on the Poor*, London: IIED.

Martínez, J. (2007), Payment for environmental services in Mexico, paper presented at a side event of the 26th session of the Subsidiary Bodies of the United Nations Framework Convention on Climate Change, Bonn, Germany, 11 May.

Miranda, M., I.T. Porras and M.L. Moreno (2003), *The Social Impacts of Payments for Environmental Services in Costa Rica: A Quantitative Field Survey and Analysis of Virilla Watershed*, London: IIED.

Muñoz-Piña, C., A. Guevara, J.M. Torres and J. Brana (2008), Paying for the hydrological services of Mexico's forests: analysis, negotiations and results', *Ecological Economics* **65**, (4)725–36.

Pagiola, S. (2008), Payments for environmental services in Costa Rica, *Ecological Economics* **65**, 712–24.

Pagiola, S., A. Arcenas and G. Platais (2005), Can payments for environmental services help reduce poverty? An exploration of the issue and the evidence to date from Latin America', *World Development* **33** (2), 237–53.

Ross, M., B. Depro and S. Pattanayak (forthcoming), Assessing the Economy-wide Effects of the PSA Programme, in G. Platais and S. Pagiola (eds), *Ecomarkets: Costa Rica's Experience with Payments for Environmental Services*, Washington DC: World Bank.

Sánchez-Azofeifa, G., A. Pfaff, J.A. Robalino and J.P. Boomhower (2007), Costa Rica's Payment for Environmental Services Programme: intention, implementation, and impact', *Conservation Biology* **21** (5), 1165–73.

Sierra, R. and E. Russman (2006), On the efficiency of environmental service payments: a forest conservation assessment in the Osa Peninsula, Costa Rica, *Ecological Economics* **59**, 131–41.

Tattenbach, F., G. Obando and J. Rodríguez (forthcoming), Generación de Servicios Ambientales, in G. Platais and S. Pagiola (eds), *Ecomarkets: Costa Rica's Experience with Payments for Environmental Services*, Washington DC: World Bank.

Thomas, V. (1998), Knowledge, Finance and Sustainable Development, in I. Serageldin, T. Husain, J. Martin-Brown, G.L. Ospina, and J. Damlamian (eds), *Organizing Knowledge for Environmentally and Socially Sustainable Development: Proceedings of a Concurrent Meeting of the Fifth Annual World Bank Conference on Environmentally and Socially Sustainable Development 'Partnerships for Partnerships for Global Ecosystem Management: Science Economics and Law'*, Washington DC: World Bank.

World Bank (2008), World Bank Projects and Operations Database, available at http://web.worldbank.org/WBSITE/EXTERNAL/PROJECTS/0,,menuPK:11 5635~pagePK:64020917~piPK:64021009~theSitePK:40941,00.html (accessed 21 September 2008).

World Bank (2007), Implementation completion and results report (Loan No. 4557, GEF Grant No. TF-23681) On a Loan in the amount of US US$32.6 million and a Global Environment Facility Trust Fund grant in the amount of US$8.0 million to the Republic of Costa Rica for the Ecomarkets Project February 9, 2007. Report No: ICR0000433, Washington DC: World Bank.

World Bank (2006), Project Document on a Proposed Loan in the Amount of USD 45.00 Millions and proposed Grant from the Global Environmental Facility Trust Fund in the Amount of 15.00 Million to the United Mexican State for an Environmental Services Project. Report No. 33228-MX, Washington DC: World Bank.

World Bank (2000), Costa Rica: Ecomarkets Project, Project Information Document PID 8876,Washington DC: World Bank.

World Bank Carbon Fund Unit (2008), BioCarbon Fund Project Portfolio, available at http://carbonfinance.org/Router.cfm?Page=BioCFandFID=9708andIte mID=9708andft=ProjectsT1 (accessed 2 October 2008).

Wunder, S., S. Engel and S. Pagiola (2008), Taking stock: a comparative analysis of payments for environmental services programmes in developed and developing countries, *Ecological Economics* **65**, 834–52.

Wunder, S. and S. Wertz-Kanounnikoff (2008), Payments for environmental services: guidance paper for the Scientific and Technical Advisory Panel (STAP), Washington DC: GEF Secretariat.

Zbinden, S. and D.R. Lee (2005), Paying for environmental services: an analysis of participation in Costa Rica's PES programme', *World Development* **33** (2), 255–72.

APPENDIX 2.1   GEF PROJECTS WITH A PES COMPONENT

Table A2.1   GEF projects with a PES component

Country	Status	Project Title (GEF ID #)
AFRICA		
Kenya	A: 2007	Agricultural Productivity and Sustainable Land Management (2355)
South Africa	A: 2004	C.A.P.E. (Cape People and the Environment) Biodiversity Conservation and Sustainable Development Project (1516)
South Africa	A: 2007	National Grasslands Initiative (2615)
South Africa and Lesotho	A: 2001	Maloti-Drakensberg Conservation and Development Project (762)
Uganda	A: 2008	Developing an Experimental Methodology for Testing the Effectiveness of PES to Enhance Conservation in Productive Landscape in Uganda (3628)
LATIN AMERICA		
Bolivia	A: 2002 C: 2005	Removing Obstacles to Direct Private-Sector participation in In-Situ Biodiversity Conservation (PROMETA) (1794)
Brazil	A: 2005	Sustainable Land Management in the Semi-Arid Sertao (2373)
Brazil	A: 2005	Ecosystem Restoration of Riparian Forests in São Paulo (2356)
Brazil	A: 2007	Espirito Santo Biodiversity and Watershed Conservation and Restoration Project (2764)
Brazil	A: 2005	Rio de Janeiro Integrated Ecosystem Management in Production Landscapes of the North-North-western Fluminense (1544)
Colombia	A: 2006	Colombian National Protected Areas Conservation Trust Fund
Colombia, Costa Rica, Nicaragua	A: 2002 C: 2008	Integrated Silvo-Pastoral Approaches to Ecosystem Management (947)
Colombia, Ecuador, Peru, Venezuela	A: 2006	Conservation of the Biodiversity of the Paramo in the Northern and Central Andes (1918)
Costa Rica	A: 2001 C: 2005	Ecomarkets (671)

Table A2.1   (continued)

Country	Status	Project Title (GEF ID #)
Costa Rica	A: 2006	Mainstreaming Market-based Instruments for Environmental Management Project (2884)
Dominican Republic	A: 2006	Demonstrating Sustainable Land Management in the Upper Sabana Yegua Watershed System (2512)
Ecuador	A: 2008	Management of Chimborazo's Natural Resources (3266)
Mexico	A: 2006	Environmental Services Project (2443)
Panama	A: 2006	Second Rural Poverty, Natural Resources Management and Consolidation of the Mesoamerican Biological Corridor Project (2102)
Peru	A: 2007	Strengthening Biodiversity Conservation through the National Protected Areas Program (2693)
Venezuela	A: 2006	Biodiversity Conservation in the Productive Landscape of the Venezuelan Andes 2120)
GLOBAL		
Worldwide	A: 2007	Institutionalizing Payments for Ecosystem Services (2589)

Notes: A=Approved; C=Closed

3. Mexico's PES-carbon programme: a preliminary assessment and impacts on rural livelihoods[1]

Esteve Corbera

INTRODUCTION

Mexico's carbon payments programme was established in 2004 as a component of a wider programme of Payments for Carbon, Biodiversity and Agroforestry Services (PSA-CABSA) which aims to provide financial incentives to rural communities and private landowners for the design and implementation of carbon sequestration, biodiversity conservation and agroforestry projects (Corbera 2005). PSA-CABSA complemented a programme of Payments for Watershed Services (PSA-H), which was established in 2003 (Alix-Garcia et al. 2005). In 2006, both PSA-CABSA and PSA-H were merged into a single policy framework of Payments for Environmental Services, with each sub-programme (hydrological, biodiversity, carbon and agroforestry services) maintaining its own procedural rules.[2] During the first two years of PSA-CABSA, the Mexican Government was the only buyer of carbon offsets, and funded rural communities and private landowners to develop projects with the technical support of consultants and civil organizations. However, government and service providers' rights and duties have changed over time due to ongoing procedural learning and variable levels of public funding.

This chapter examines the design and performance of Mexico's PSA-CABSA programme, with an emphasis on its carbon component. Methodologically, it relies on secondary literature and interviews with key stakeholders, including government officials, consultants and community leaders, conducted over two research periods (January–April 2007 and October–November 2008). The second section examines the design of PSA-CABSA and the rules for carbon payments, paying attention to the design process, geographical coverage, actors' rights and responsibilities, level of compliance, sanctions and the price-setting mechanism for carbon offsets. The third part of the chapter addresses the impacts on participants' natural, human, physical, financial and social capitals and the conclusion illustrates how PES programmes may inform the development of REDD activities in Mexico, and outlines a future research agenda on PES.

THE DESIGN OF MEXICO'S PSA-CABSA AND THE CARBON PROGRAMME

According to the country's third National Communication on Climate Change, 64 484 kilotons of carbon dioxide equivalent were emitted from forest conversion over the period 1990–2002, representing approximately 10 per cent of the country's overall emissions for that same period (Comisión Intersecretarial sobre Cambio Climático 2006). Deforestation is caused by forest fires, the government's failure to regulate private and state-led logging, as well as rural communities' clandestine woodcutting (Klooster 1999) and in-migration by dispossessed peasants to forest areas (O'Brien 1998; de Vos 2002). Some scholars also argue that deforestation has been caused by rural communities transforming forests into pastures and agricultural lands, particularly where suitable geophysical characteristics are present, and collective conservationist behaviour is weak (Alix-Garcia et al. 2005).

About 5 per cent of forest lands are state property, 15 per cent are privately owned, and 80 per cent are owned by indigenous and other local communities who practice agriculture and forest management on household lands or common forests, known as *ejidos*[3] (see Chapter 1). Although the government only started supporting community forestry in the early 1980s, it is evident that any effort to halt deforestation successfully requires involving peasant communities. Management plans have been developed for 25 per cent of communal forests (Klooster 2003) and, specifically in highland areas, forest recovery is occurring as a result of out-migration to urban centres and abroad (Rudel 2008).

The National Forestry Commission (CONAFOR) believed that providing incentives for the provision of environmental services could halt deforestation and reinforce community-based forest conservation and sustainable forest management (Klooster and Ambinakudige 2005). The General Law for Sustainable Forest Development, passed in February 2003, established the Mexican Forestry Fund as a financial instrument to promote the establishment of incentive- and market-based systems for the conservation of forest ecosystems, to be managed and administered by CONAFOR. Furthermore, a modification of Article 223 in the Law of Rights established that a small levy on water tax payments, managed by the National Water Commission (CNA, in Spanish), should be channelled to the forestry fund in order to develop the PSA-H programme and support the conservation of forest environmental services (Muñoz-Piña et al. 2008). In late 2006, CONAFOR negotiated two grants with the Global Environment Facility and the World Bank for US$15 and US$45 million. The aim of these grants was to improve outreach and evaluation capacities in PES programmes, increase beneficiary numbers and promote a more effective and efficient commercialization of environmental services by strengthening monitoring systems and establishing well-functioning markets, particularly for watershed services.

Politics of Design and PES Definition

PSA-CABSA resulted from lobbying by peasant and forest-based organizations, including the Mexican Council for Sustainable Agroforestry, the Mexican Network of Forestry Organizations, the National Network of Coffee Producers Organizations and the National Union of Community Forestry Organizations, among others. During 2002 and early 2003, these organizations negotiated a National Rural Agreement with the government, which outlined a development plan for the Mexican countryside. They pushed to include a number of strategic sectors for policy development and funding, including 'a policy programme to implement payments for environmental services: (1) on carbon fixation by forests to halt climate change; (2) for rural communities who support biodiversity conservation; and (3) for the development or improvement of agroforestry systems, specifically for shade-grown coffee plantations' (Government of Mexico 2003, p. 37). These organizations believed that payments for these three activities would contribute to valuing forest conservation and sustainable management, and would lead to linking service buyers and providers through the creation of markets for environmental services.

The design phase of PSA-CABSA was characterized by competing views about the likely operational feasibility of such programmes in the short term, mostly because sustainable funding mechanisms were not easily identifiable. Government officials considered that it was more convenient to design such programmes once the PSA-H was fully operational in order to learn from its design and implementation experience. In contrast, rural and forest organizations prioritized the establishment of multiple PES schemes, however loosely defined, in order to secure resources in the short term, allowing for procedural changes as PES implementation proceeded.

The definition of an environmental service under PSA-CABSA was influenced by the preferences of the actors involved in institutional design – especially by the interests of forestry and coffee producers' organizations which supported the development of agroforestry systems as a means of conserving biodiversity and diversifying rural livelihoods. This led to the labelling of carbon sequestration, biodiversity conservation and agroforestry systems as environmental services when, in fact, they were constitutively different. According to the Millennium Ecosystem Assessment (2005), environmental services are the benefits people obtain from ecosystems, among which it is possible to distinguish supporting, provisioning, regulating and cultural services. Under this framework, biodiversity is considered an element from which some of these services are generated. Therefore, when PSA-CABSA equated payments for carbon sequestration with payments for biodiversity conservation and for the establishment of agroforestry systems, it ignored the fact that the former is a regulating service which can be accounted for (and traded) in tons of carbon, while the latter are *proxies* for the provision of a broader range of ecosystem services, such as non-timber forest products, soil conservation, genetic conservation and aesthetic values, among others, which involve distinct accounting and evaluation methods.

Evolving Rules

The original PSA-CABSA rules had a common procedural framework for its three sub-programmes (carbon, biodiversity and agroforestry), including specific rules for project design and implementation. Rural communities, landowners' associations and single landowners were regarded as service providers and had to go through a two-stage application process (see Figure 3.1). First, they applied to CONAFOR for assistance in project design, unless they already had a project ready for implementation. Applicants could not receive any funding from other PES programmes, had to prove they owned the land where the project was going to be developed, and had to prove the existence of a forest management plan, an environmental management unit or a commitment to the project through a letter written and approved by the community assembly[4] in the case of community forests. Applicants also had to demonstrate that PES activities would not have been carried out without payments (in other words, they were *additional*). Under the carbon programme, the project area had to be located within a deforested or degraded area of the country, as defined by CONAFOR in a series of publicly available maps. In the case of biodiversity, eligible areas encompassed municipalities located in regions of critical importance for watershed protection and biodiversity conservation (for example, protected areas, biological corridors and high biodiversity hotspots), while for agroforestry, prioritized areas included municipalities with shade-grown coffee production and high biodiversity.

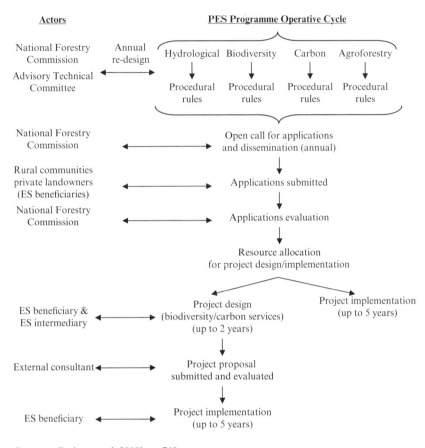

Source: Corbera et al. 2009b, p. 749.

Figure 3.1   PSA-CABSA application process

If successful, applicants received up to US$28 775 for project design, to be conducted within 2 years, with the support of a civil organization (for example, NGOs, consultants or technical organizations) acting as a service intermediary. For carbon, the project needed to define and address at least the following issues:

- the development of a baseline of current land use cover or a defor-estation risk map of the area where the project would develop;
- a description of which forest conservation and/or reforestation activities would be conducted, on which areas, and in how many hectares;

- when planting would take place and by whom, and where seedlings would come from (in the case of reforestation activities);
- the quantity of carbon emissions avoided or sequestered over a 5-year period using sound modelling techniques;
- how the activities and enhanced carbon stocks would be maintained over time, in order to ensure permanence and avoid leakage; and
- how carbon payments would be spread amongst participants, how they would be invested and how potential conflicts would be dealt with – particularly if the project involved a whole community.

Once the project design phase concluded, CONAFOR sent the project documentation to independent consultants who assessed whether project proposals complied with these criteria and would be eligible for implementation.

In 2006, PSA-H and PSA-CABSA were integrated into a single PES policy framework which fell under a broader policy framework encompassing all forestry government programmes, known as *Pro-Árbol*. The application process and eligibility criteria did not change much except for contractual flexibility rules. In 2004 and 2005, failure to comply translated only into payment cancellation, but the 2006 rules obliged participants to return payments in the event of non-compliance. This was tempered by ruling out sanctions if participants showed that their failure to comply was due to an uncontrollable reason. Resource users' rights to a fair monitoring process were secured but no sanctions have been imposed to date. Limited sanctioning is explained by the lack of human and financial resources to ensure widespread and standardized monitoring practices by service intermediaries and CONAFOR's extension officers, but also in some cases by the beneficiaries' unwillingness to authorize in-depth evaluations of project sites (Martínez Tenorio et al. 2007). It has been noted that there is a need to develop sound monitoring methods for the assessment of multi-component and complex agroforestry and biodiversity projects (Gómez Guerrero et al. 2006; Martínez Tenorio 2007).

Rules for PES-carbon projects have changed substantially since 2004. As noted above, during 2004 and 2005, eligible land-use activities included conservation, reforestation or afforestation activities, as far as such activities proved to annually sequester or avoid the release of 4000–8000 tons of carbon dioxide equivalent (tCO_2e)[5] per year or a maximum of 40000 tCO_2e over 5 years. The government – acting as the only buyer of carbon – paid between US$3.61/$tCO_2e$ and US$7.07/$tCO_2e$ only over 5 years. Payments were made annually in cash, with the amount of the final payment dependent upon the actual amount of sequestered carbon, based on monitoring by government officers. The price paid per ton was higher if the project met a number of stated criteria (for instance, being located in a protected area buffer zone, included endangered species or applicants belonged to an ethnic group with a high level of social marginalization). Project applicants to carbon funding could also be granted up to US$10 833 annually for verification of project activities, to be undertaken by an external consultant or community members; another US$10 833 for capacity-building activities (to be undertaken by an external party approved by CONAFOR); and up to US$18 060 for technical assistance and project follow-up, to be provided by community members or an external consultant, as decided by the applicants.

In 2006, procedures for PES-carbon projects evolved to make them comply with small-scale afforestation and reforestation projects under the Clean Development Mechanism (CDM), as the government had an interest in supporting projects which may potentially become sources of tradable Certified Emission Reductions (CERs)[6] after the five years of government funding ended. This would require landowners and communities to partner with service intermediaries to register and secure sales of reduced emissions. Under these 2006 rules, a minimum area for projects was set at 500 ha (and a maximum of 3000 ha for the rules of 2007), with no tree cover since 1990, and a maximum annual sequestration rate of 8000 tCO_2e. Funding for implementation was set at 80 times the Mexican minimum salary wage per hectare, for a period of 5 years, regardless of the expected amount of carbon sequestered. Mexico's minimum average salary wage in 2007 and 2008 was approximately US$3.61/day, which means the average payment became US$288/ha.

In 2008, the rules changed again and the annual sequestration threshold increased to 16 000 tCO_2e, in line with the decision made by the Parties to the Kyoto Protocol in 2007. Existing carbon 'sellers' continued to be bound to previous rules, while new applicants received funding to develop a CDM Project Idea Note (PIN), following the guidelines of the Kyoto Protocol's mechanism (see www.unfccc.int). CONAFOR funding to develop a PIN was provided where there was a proven record of the service intermediary partnering with the applicants and according to the presence of key project characteristics, including its location in a national protected area, an important ecoregion or mountainous range. Applicants receive 50 per cent of the funding at the start of PIN design and the remaining 50 per cent once the PIN has been submitted and approved by CONAFOR.

These changes highlight the ad hoc definition (independent of international carbon markets) of the economic value of carbon by CONAFOR. In contrast, for hydrological services, CONAFOR worked with the scientific support of the National Institute of Ecology (INE) to define PSA-H payments. A study of profits from agriculture and livestock operations near forested areas (Jaramillo 2002 cited in Muñoz-Piña et al. 2008) helped to define the distribution of opportunity costs of conserving forests as an input to the design of PSA-H. The study shows average annual profits of US$37/ha from growing corn, and of US$66/ha for livestock production. The estimated distribution showed that with a payment of US$18.20 per hectare, more than 40 per cent of forest owners in the sample would have preferred conserving forests to converting them to cornfields. The same payment would have stopped 12 per cent of pasture owners from deforesting their land. Based on this analysis, initial PSA-H payments were set at US$18.20/ha, except for cloud forests, which were set at US$27.30/ha. Nevertheless, Muñoz et al. (2008) recognized that this fixed-price payment system was sub-optimal, as it did not differentiate between those with lower opportunity costs and those with higher costs. INE thus developed a national deforestation risk map to rank PSA-H applications from 2007 onwards, which assumes that the risk of deforestation is a proxy for opportunity costs related to urbanization, conversion to agriculture (mainly corn), logging and cattle ranching (particularly in tropical areas); the lower the risk, the lower the opportunity cost. This has enabled INE and CONAFOR to overlay PSA-H applications onto deforestation risk maps and differentiate payments accordingly.

Alongside the procedural changes, CONAFOR set up a Technical Advisory Council (TAC) formed by ecosystem service intermediaries, academic institutions, civil organizations and government departments, and organized around several working groups. Regular meetings have been held since its creation in September 2006. The TAC's role is to advise CONAFOR on issues including how to improve PES rules; criteria for PES eligibility areas; how to evaluate project developers' performance; key lessons derived from PES external evaluation reports; the economic value of environmental services; forest policy integration and perverse incentives and how to expand PES programmes and establish real markets for ecosystem services (Bezaury-Creel 2007). Recently, the TAC has also been involved in discussions about the procedures for a Mexican REDD framework under the auspices of the World Bank Forest Carbon Partnership.

Geographical Coverage and Project Design Issues

Between 2004 and December 2007, projects covering over 1 million hectares received support to design and implement PSA-CABSA projects (see Table 3.1). The reduction in the number of hectares receiving funding for design and implementation in 2005 and 2006 can be explained by the considerable reduction in the level of funding in those years – in 2004

Table 3.1   Annual and total hectares under PSA-CABSA, 2004–07[a]

Year	Project design[b] (ha)	Project implementation[c] (ha)	Total (ha)
2004	526225	31448	557673
2005	20520	29477	49997
2006	56013	7281	63294
2007	291657	64835	356492
Total			1027456

Notes:
[a] Excludes Payments for Hydrological Services.
[b] Projects at their design stage, which have not yet been independently reviewed and have not received funding for implementation.
[c] Projects already receiving support for implementation.

Source: Corbera et al. 2009b, p. 752 and updated for CONAFOR 2007 data.

funding was approximately US$7 million, in 2005 it was approximately US$4 million, dropping to about US$0.83 million in 2006. Over the period 2004–06, PSA-CABSA funding came from contributions made by the CNA and the National Commission for the Development of Indigenous Peoples (CDI, in Spanish). In 2007, funding increased after a 2-year negotiation between CONAFOR and the Ministry of Finance, which finally led the Mexican Congress to provide CONAFOR with at least US$10 million/year for PSA-CABSA implementation (in 2008, this contribution increased further to US$18 million), in addition to the existing variable contributions from CNA and CDI.

The expenditure per hectare over the period 2004–06 for project implementation fell from US$121/ha to US$54/ha, and remained relatively stable for project design, peaking in 2005. In 2007, both ratios increased again. Funding has been unevenly distributed across the country – between 2004 and 2007 Oaxaca, Chiapas, Durango and Veracruz received more than 50 per cent of total PSA-CABSA funding, explained by the higher number of applications from these states, the better quality of the submitted proposals, and the support received to prepare proposals from service intermediaries and NGOs, which are more numerous in these regions (Martínez Tenorio et al. 2007).

The number of PSA-CABSA applications remained steady between 2004 and 2007 (Corbera et al. 2009a), suggesting that it has been attractive for rural communities. Between 2004 and 2006, public funding was insufficient to support the total number of successful applications and 5–10 per cent of applications had to be deferred until more funding became available. Nevertheless, rejection rates of applications and project design documents have been relatively high across all sub-programmes, due to factors including missing documentation, non-fulfilment of eligibility criteria, and lack of environmental additionality in the case of carbon projects.

An external evaluation of 32 out of 87 carbon project proposals submitted in 2005 and 2006 concluded that only four projects could be approved to proceed with implementation (Ruiz 2007), which represents a relatively small proportion of applicants. Seventeen of these 32 projects were unable to establish an adequate baseline scenario or to account for the possibility of inducing land-use change somewhere else as a result of the project (in other words, negative leakage). In some cases, comprehensive carbon inventories were prepared but the project proposals were weak. In others, projects did not meet key eligibility criteria (for example, the maximum annual sequestration rate of $8000\,tCO_2e$ for small-scale CDM projects) or they confused terms such as coal or charcoal with carbon dioxide. Finally, at least two documents contained nothing but a full site description. Successful projects employed carbon accounting methodologies validated by the Intergovernmental Panel on Climate Change and the World Bank's BioCarbon Fund, based on biomass calculations for different vegetation and soil strata, and relied on bibliographical references, satellite imagery and on-site sampling with local community participation (Martínez Tenorio et al. 2007).

Analysis of project approval rates by Corbera et al. (2009b) suggests that considerable funding was wasted in preparing unsuccessful project proposals, as project developers lacked the necessary knowledge and capacity to design them. This was in turn related to the constraints of CONAFOR staff in communicating the principles of PES projects to service providers and intermediaries. This is supported by a statement from one staff member:

> The lack of technical capacities has been the most important barrier for [successful] project design. There are very few forestry engineers out there who can develop viable carbon projects. Moreover, there has been a lack of coordination between forestry consultants and the rural communities. In most cases, they have not properly coordinated and this is why some projects will never move from design to implementation. Sometimes, consultants design a project which is not fully supported by local communities. In other cases, communities develop activities without the support of their consultant. (Environmental consultant interview, October 2008)

This highlights the need to guarantee the rights of local communities to prosecute service intermediaries who mismanage funds or who do not deliver well-designed project proposals, for example through formal contracting supervised by CONAFOR or an independent legal entity.

It must also be acknowledged, however, that rural communities and service intermediaries have been on a steep learning curve in recent years in Mexico. While the average rejection rate over the period 2004–06 was above 80 per cent, it dropped below 50 per cent in 2007, suggesting that the number of applications missing information or failing to meet eligibility criteria had declined substantially. This was related to CONAFOR's efforts to improve the technical capacity of service intermediaries, a critical aspect acknowledged in programme evaluations (Gómez Guerrero et al. 2006; Martínez Tenorio et al. 2007). *Pro-Árbol* procedural rules for 2007 and 2008 established that service intermediaries had to undertake a capacity-assessment process before providing technical assistance to rural communities interested in PES and other forestry programmes. In 2008, CONAFOR employed 47 professionals to assist communities with preparing project proposals, to communicate rules to potential beneficiaries and to oversee implementation across the country.

There has been a variable success rate with carbon applications, with a rejection rate of between 55–97 per cent between 2004 and 2007 (Martinez Tenorio et al. 2007; Corbera et al. 2009b). One of the projects which received funding for implementation in the early years is currently commercializing (brokering and selling) voluntary carbon offsets under an NGO-led initiative known as *Neutralízate* (that is, becoming carbon neutral). Offsets are sold at US\$10 per tCO_2e, and 80 per cent of this value reaches service providers. In most other cases, however, it appears that too many resources have been allocated to projects which may find it difficult to sell any carbon offsets in international markets – in part because forestry projects have lost appeal in voluntary carbon markets over recent years in favour of renewable energy projects and there appears to already be a higher level of supply from voluntary, standard-based, carbon forestry projects than there is demand in the market (Corbera et al. 2009a). It also remains to be seen how many of the CDM-like projects will make it through the validation and registration phases of this mechanism. In this context, the viability of government-led carbon forestry projects in Mexico, or at least the future development of their financing stream, will be subject to service providers and intermediaries' ability to capture the attention of investors in international carbon markets, which in turn may be influenced by post-2012 Kyoto negotiations, the role of forestry projects under such a framework and public perceptions on forestry offsets (Boyd et al. 2008).

Implementation, Transaction and Opportunity Costs

Opportunity and transaction costs, such as programme outreach, monitoring and evaluation are under-researched aspects of PES programmes. CONAFOR used US$0.7 million to cover transaction costs over the period 2004–2007. This represents over 3 per cent of the overall budget, falling below the 4 per cent threshold established by law to implement and monitor PSA-CABSA. Government-led PES programme regulations, which keep transaction costs under a certain threshold, can make such estimates misleading, as activities whose costs do not fit within this cap may be postponed or undertaken at lower than optimal levels if external funding cannot be found to pay for them and, conversely, a decision to increase payments can automatically expand transaction budgets in direct proportion to the increase (Wunder et al. 2008, p. 848).

In the latest public evaluation of PSA-CABSA, Martínez Tenorio et al. (2007) interviewed 232 individuals and representatives from communities participating in PSA-CABSA and representing 68 per cent of all the beneficiaries receiving payments for project implementation.[7] This study showed that the only land-use activity which is less profitable than PES payments (per hectare) was traditional maize cultivation and that any other activity, such as cultivation of beans and chilli, or grazing, would be more beneficial to the landowner (Martínez Tenorio et al. 2007). The study also provided a more general assessment of the relationship between economic benefits and costs for each PSA-CABSA programme (agroforestry, biodiversity and carbon payments). They indicated that implementation costs varied significantly across programmes due to divergent time investment by households, payments for external labour, project area size and types of management activities undertaken. Comparing average costs and payments, the authors concluded that the best 'economic benefit versus cost' ratio was associated with agroforestry projects, followed by carbon projects, if technical assistance for carbon accounting and monitoring was covered by CONAFOR. Biodiversity projects often did not cover their costs, as they involved more labour, infrastructure development (for instance, hides and wells for fauna, road improvement) and additional assistance from experts in fauna and flora.

Programme evaluations do not provide a detailed assessment of PES payments and opportunity costs for each participant, or consider willingness to participate in programmes. Such analyses may not be pertinent in carbon and biodiversity projects, where activities are implemented in forest commons, which are often already set aside under community regulations for the provision of timber, fuelwood, grazing and non-timber forest products, thus reducing the importance of opportunity costs in

Table 3.2   Case studies analysed

Community/ Ejido	Population	House-holds	Land area (ha)	Project-design start date	Project-implementation start date
San Bartolomé Loxicha	1706	522	14 076	2004	2005
Orilla del Monte	2353	546	2 764	2005	2006
Oyamecalco El Cajón	479	97	1 200	2005	2006
Niños Héroes de Chapultepec	141	33	1 800	Does not apply	2005

collective decision making. In this context, willingness to participate is more strongly influenced by other variables such as group size and diversity, population dynamics and conservation values (Kosoy et al. 2009).

In the case of agroforestry, eligible areas frequently overlap with areas where shade-grown coffee cultivation is already practised. Thus, most participants receive payments to improve their cultivation systems through the diversification of shade-tree species and the development of live fences, which do not imply a costly conversion from one land-use type to another. This compensatory (rather than transformative) character of PSA-CABSA has also been experienced in the PSA-H programme, where high-risk deforestation areas with high opportunity costs were not targeted during the first 3 years of implementation (Muñoz-Piña et al. 2008). The key question facing Mexican PES programmes is thus whether they should only target areas characterized by high deforestation risk and more profitable, commercial land-use activities so as to maximize efficiency, or instead compensate for a less efficient production of environmental services with lower environmental additionality.

IMPACTS ON RURAL LIVELIHOODS

As PSA-CABSA projects have been operational for only a few years, and hundreds of projects are still in their design phase, it is not possible to provide conclusive data on livelihood impacts. Therefore, what follows provides a preliminary assessment according to the findings of independent evaluations commissioned by CONAFOR (Gómez Guerrero et al. 2006; Martínez Tenorio et al. 2007) and fieldwork developed in four out of seven PES-carbon projects being implemented in 2007 and described in Table 3.2 and Figure 3.2 (Corbera et al. 2009b). In contrast with programme

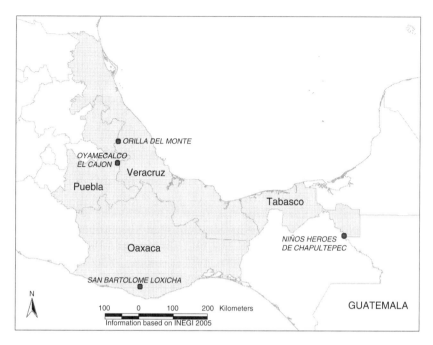

ORILLA DEL MONTE

OYAMECALCO
EL CAJON

Veracruz

Puebla

Tabasco

NIÑOS HEROES
DE CHAPULTEPEC

Oaxaca

SAN BARTOLOME LOXICHA

N

100   0   100   200 Kilometers

Information based on INEGI 2005

GUATEMALA

Source: Corbera et al. 2009b, p. 749.

Figure 3.2   Location of the four carbon projects analysed

evaluations, this research relies on direct insights from PES-carbon participants. Focus groups of between 18 and 53 people were conducted to document participants' perceptions of carbon payments, involving local authorities, farmers and, in some cases, CONAFOR extension officers and service intermediaries. Research objectives were explained and a series of graphic materials were used to elucidate farmers' views regarding: why they joined the project; what benefits they had derived from participation; how they organized themselves to implement the project; which activities they were currently developing; whether any conflicts had arisen as a result of project implementation; and which aspects of the project should be improved.

Natural Capital

The first and principal objective of PSA-CABSA concerns the conservation and enhancement of the biological resources available to rural communities and the maintenance of a diverse resource portfolio, for instance, by increasing the number of standing trees and promoting biodiversity conservation and sustainable land-use management activities. By the end of 2007, 133 041 ha of land was under the PSA-CABSA programme, which could increase to over 1 000 000 ha over the next few years if all the projects currently in the pipeline receive funds for implementation. Of this, the seven carbon projects under implementation through PSA-CABSA between 2004–06 encompass 11 147 ha and expect to sequester 220 567 tCO$_2$e over a period of 25–30 years (Martínez Tenorio et al. 2007).

Martínez Tenorio et al. (2007) found that over 70 per cent of their interviewees showed a willingness to preserve the land under PSA-CABSA from other land uses such as agriculture or grazing activities regardless of whether or not PES incentives were received, which demonstrates that projects often tend to reinforce existing positive attitudes towards environmental conservation. The sequestration rates of the four case studies analysed range between 20 064 tCO$_2$e to 32 291 tCO$_2$e in the first 5 years of implementation. This translates into divergent sequestration rates per hectare, due to the different ecosystems involved and the different methodologies used for calculation.

Interviewees in *San Bartolomé Loxicha* considered the programme as an opportunity to support ongoing reforestation and conservation activities in the local commons and perceived that reforestation activities with *Pinus* sp. would increase the amount of water in their springs and that forest conservation would be important to have access to 'clean air'. In *Orilla del Monte*, the project supported the planting of *Pinus cembroides* on households' properties, in order to commercialize pine nuts in the future, demonstrating how a PES-carbon project can contribute to the incorporation of new agricultural activities, whilst increasing the number of future potential sources of financial capital, as described below.

In the other two *ejidos*, planted trees helped to increase tree stocks in the commons and individual agricultural plots, but there were also divergent views regarding the project's impact on natural capital. In *Niños Héroes*, for instance, several farmers argued that the project was an impediment to increasing the amount of land dedicated to agriculture, although they recognized that payments were a good strategy to increase collective and household income. Farmers reforested 843 ha with *Tabebuia rosea* and carried out conservation, fire control and monitoring activities both in the forest commons and on household property. Farmers in *Oyamecalco El Cajón* planted a locally adapted *Pinus pseudostrobus*, which has increased interest in controlling livestock access to the commons in order to preserve the newly planted trees.

Human Capital

Human capital refers to the labour available to the household, as well its education, skills and health (Ellis 2000). In this regard, there are no detailed analyses in Mexico or elsewhere of how PES programmes and projects impact upon participants' labour and the amount of time each participant allocates to project implementation. There is evidence, however, that projects contribute substantially to participants' increased awareness of conservation issues and ecosystem services, in particular, as well as to strengthening forest management skills (Gómez Guerrero et al. 2006; Martínez Tenorio et al. 2007).

The projects examined show that carbon payments can enhance human capital. In *San Bartolomé Loxicha*, the community assembly created a forest committee to coordinate planting activities and its members benefited from a forest management training course provided by CONAFOR. Nevertheless, interviewees also felt that there was a need to receive more technical support from CONAFOR to apply for new funding opportunities and prepare complementary projects. In *Orilla del Monte*, a group of 95 farmers established a committee to coordinate project activities and facilitate information flows between CONAFOR, the *ejido* assembly and project developers. This illustrates how PES increased project and resource management skills and prompted the creation of new organizations within participant communities, a form of social capital.

PSA-CABSA has also contributed to new skills among service intermediaries, including the NGOs, consultants and technical organizations helping rural communities and farmers to apply for funding and design the projects. Programme evaluations and the focus groups showed a general satisfaction with the intermediaries' role and involvement in project development and local capacity-building. However, while intermediaries' involvement in the programme has generally been positive (insofar as they have maximized rural communities' participation and probably decreased the application rejection rates), it has been acknowledged that intermediaries have sometimes influenced the objectives and management strategies too much, undermining rural communities' views (Martínez Tenorio et al. 2007). Furthermore, some intermediaries have not been competent, leading to conflict between intermediaries and communities regarding project design and how funding should be shared among parties (Corbera et al. 2009b).

In Mexico, where rural poverty and ethnicity are positively correlated (Patrinos and Skoufias 2007), it is worth highlighting the extent to which PSA-CABSA has benefited indigenous populations, making an effort to provide more opportunities to the most marginalized. In this regard, the proportion of PSA-CABSA projects implemented by indigenous communities in their own territories has risen from 12 per cent in 2004 to 31 per cent in 2006 (Martínez Tenorio et al. 2007). However, the number of women beneficiaries has never risen above 6 per cent of total beneficiaries. This can be explained, first, by the fact that women own less than 25 per cent of formalized property, including private and household property within *ejidos* and indigenous communities, though they represent over 50 per cent of Mexico's rural population (Martinez Tenorio et al. p. 43). Second, Mexico's community-based institutions are typically controlled by men, and therefore so are decisions concerning forest management (Corbera et al. 2007).

Physical Capital

PSA-CABSA has not been explicitly designed to support rural communities and individual landholders to develop or maintain physical capital and, consequently, programme reviews do not provide quantitative or qualitative evidence in this respect. It is acknowledged that some biodiversity projects under implementation have improved local roads and built patrolling and control plots in conservation areas (Martínez Tenorio et al. 2007). These case studies provide minimal anecdotal evidence in this respect. The *ejido Oyamecalco El Cajón* bought a radio and a computer for the community, whilst *Niños Héroes* invested a share of payments to fix a water pump which had been broken for some time. It also built a greenhouse to produce seedlings for the project. In the literature on carbon forestry projects, there is little evidence of carbon projects' contribution to physical capital and there is only anecdotal evidence of participants investing in tools, machinery or local infrastructure. Corbera (2005) indicates that participants in a project commercializing voluntary carbon in Mexico invested a share of project revenues in agricultural tools, as well as in improving community buildings.

Financial Capital

Martínez Tenorio et al. (2007) state that more than 75 per cent of their interviewees earned less than US$362 per month,[8] and lived in poor and marginalized areas, explaining their main reasons for joining the programme as: to 'earn an additional income' (54 per cent), 'halting deforestation' (26 per cent) and 'changing land use in non-productive lands' (10 per cent). That report also showed that over 64 per cent of beneficiaries considered PES incentives to represent less than 25 per cent of their monthly income, 30 per cent noted that they comprised between 25–40 per cent, and 6 per cent stated that PES incentives made up more than half of their monthly income. These findings are consistent with a previous assessment, which indicates that more than 50 per cent of the beneficiaries surveyed stated that PES payments were not high enough to make a substantial difference in their lives (Gómez Guerrero et al. 2006). When participants were asked about the impact of the project on their well being, 30 per cent considered that it had improved their lives significantly and around two-thirds indicated that it did not make any difference (Martínez Tenorio et al. 2007, p. 53). These results suggest that, in conditions of extreme and acute rural poverty, an increase in income of 25 per cent or more allows households to increase available savings and purchasing power but does not change their relative poverty condition, if the latter is understood as the deprivation of basic capabilities, such as freedom of choice and the real opportunities one has in life (Sen 1999).

In the four case studies, the community of *San Bartolomé Loxicha* – which may generate US$221 855 in carbon payments over 5 years – had already received US$145 000 in carbon payments, and service interme-diaries had received US$28 000 for capacity-building and project follow-up. The community assembly granted 20 per cent of carbon revenues (US$29 000) to the *Milenio* Coffee Producers Organization, which brings together 100 farmers who are mostly involved in carbon project activities. The remaining 80 per cent (US$117 000) was distributed among farmers involved in tree planting. As the number of participants involved in plant-ing each year varied, it was not possible to estimate an average figure of PES-generated household income. However, if the total number of house-holds is taken into account and it is assumed that every household partici-pates in planting activities, one comes up with a rough figure of US$112/household/year.

In *Orilla del Monte*, total payments of US$136 088 were expected over the 5-year implementation period, with approximately US$20 000 received annually in 2006 and 2007. Of this, 90 per cent was distributed among those involved in tree planting on their household plots, whilst the remain-ing 10 per cent was allocated to cover management and technical expenses incurred by project developers. The focus group did not estimate average PES income per household, as each participant planted a variable number of trees on her own plot and payments differed. However, again assuming that all households planted trees on their plots and in the same quantity, average carbon payments were estimated at US$33 per household per year.

Oyamecalco El Cajón also received approximately US$20 000 in 2006 and in 2007 from the US$147 979 expected for the period 2006–10. As in *San Bartolomé Loxicha*, this *ejido* rewarded only those involved in reforestation activities and forest patrolling. Average annual payments per household of US$221 were estimated, based on the assumption that everybody participated. Finally, *Niños Héroes* received US$125 077 in carbon payments for 2005–07, representing 51 per cent of total expected payments (US$244 513). Carbon revenues were distributed among those who participated in planting and patrolling activities on their own land and the forest commons, and on fixing a water pump and building a greenhouse as indicated earlier. Interviewees in this case noted that this translated into an average payment of US$508/household/year, which is by far the highest figure of the four cases due to the smaller number of households in the *ejido* than in the other three.

Social Capital

The complexities associated with defining social capital explain why this chapter only describes the extent to which PSA-CABSA programmes, and PES-carbon in particular, have been able to strengthen and expand social relationships and networks, as well as whether and how they have contributed to generate conflict and debilitate trust and reciprocity among community members and other social actors.

Approximately 65 per cent of interviewees in programme evaluations and in our case studies acknowledged that the programme had promoted community organization, by strengthening collective action through land-use management and tree-planting activities (Martínez Tenorio 2007; Corbera et al. 2009b). As described above, projects have led to the establishment of forest management groups, which has translated into a greater commitment from participants to control forest fires, illegal tree-cutting and livestock roaming in the commons. Local field research in *Niños Héroes*, however, revealed some level of conflict resulting from the community's assembly decision to involve settlers in reforestation activities and to reward them with only half of the carbon income allocated to formal right-holders. This case is apparently an exception, but underscores the importance of property rights relationships in mediating access to natural resources, ecosystem services and their benefit streams. Table 3.3 (opposite) summarizes the impacts of the carbon projects on the five 'capitals' of the livelihood framework for the four case studies.

CONCLUSIONS

Mexico's PES programmes have been successful in attracting a large number of rural communities and individual landholders. The latest data

Table 3.3 *Carbon sequestration, project finance and social benefits in the PES carbon projects analysed, 2004–06*

Community/ Ejido	Impact on natural and financial capitals						Impact on human, physical and social capitals
	Project activities (hectares)	Funding for implementation over five years[a] (US$)	Carbon sequestration over five years (tCO$_2$e)	Sequestration per hectare (tCO$_2$e/ha)	Carbon price[b] (US$/tCO$_2$e)	Household benefit (US$/household/ year)[c]	
San Bartolomé Loxicha	Reforestation with *Pinus pseudostrobus, P. patula* and *P. oaxacana* in the forest commons (1040)	221 855	32 291	31.0	6.87	112	New forest management committee Forest management training course
Orilla del Monte	Reforestation with *Pinus cembroides* Reforestation with *Pinus cembroides* in households' land holdings (1000)	136 088	20 064	20.1	6.78	33	New forest management committee
Oyamecalco El Cajón	Reforestation with *Pinus pseudostrobus* in the forest commons (201)	147 979	21 114	10.5	7	221	Purchase of a radio and a computer

Table 3.3 (continued)

| Community/ Ejido | Impact on natural and financial capitals | | | | | Impact on human, physical and social capitals |
	Project activities (hectares)	Funding for implementation over five years[a] (US$)	Carbon sequestration over five years (tCO$_2$e)	Sequestration per hectare (tCO$_2$e/ha)	Carbon price[b] (US$/tCO$_2$e)	Household benefit (US$/household/year)[c]	
Niños Héroes de Chapultepec	Reforestation with *Tabebuia rosea* in households land holdings and the forest commons (843)	244 513	26 723	31.7	9.14	508	Water pump repair New greenhouse Conflict between formal and informal right holders

Notes:

a   This excludes funding for project design, capacity-building and technical assistance.

b   Carbon prices are estimated taking into account the total investment in the project, including funding for capacity-building or technical assistance, the number of ha under the project, and the expected sequestration rate.

c   These figures are indicative (see text).

(September 2008) show that there are over 2600 communities, associations and private right-holders receiving payments for watershed, biodiversity, carbon sequestration services and the establishment of agroforestry services, occurring on more than 1.75 million ha (Cibrián Tovar et al. 2008). However, it is still too early to evaluate the overall performance of these programmes. As this chapter has shown for the case of carbon, and to a lesser extent for agroforestry and biodiversity projects, preliminary research shows that payments have been positively received by local communities and there are positive impacts on rural livelihoods' natural, human, physical, financial and social 'capitals'. It can be argued that the improvement of natural and financial capital constitutes the key pillar of PES, which also has positive spillovers into human, physical and social assets.

It has been shown that the number of hectares of PSA-CABSA promoting active and sustainable land-use management activities are increasing over time, securing the provision of several ecosystem services. Projects' revenues are invested in collective goods, such as improving local infrastructure and institutions, forest management skills and other productive activities, but also in individual households' needs, such as improving the house, buying food staples and supporting children's education, among others. However, conflict has also resulted from misunderstandings regarding the distribution of payments among community members with divergent rights and responsibilities, and because of service intermediaries' inability to deliver good project proposals. During the early years of PSA-CABSA, in particular, many service intermediaries lacked the right skills to help communities to apply for, design and implement projects. Over time, these skills have been improved and CONAFOR has made an effort to register all intermediaries and to ensure that they are competent. This is a very positive step insofar as intermediaries contribute to building trustworthy relations between providers, CONAFOR and themselves.

These service intermediaries may also have to play a leading role in establishing long-term financial mechanisms and monitoring programmes for PES schemes, in order to support effective ecosystem conservation and environmental service provision (Martin et al. 2008). In fact, in the carbon projects analysed above, the participants' main worry concerned the duration of CONAFOR's financial support; they argued that such support should continue after 5 years if they show proof of good organization and tangible conservation results. Existing programmes such as Mexico's PSA-H or Costa Rica's PSA have addressed this financial challenge by earmarking federal fees on water tariffs, fuel taxes or both, and the programmes coexist with local initiatives in which contracts between environmental services consumers and providers have been established

(Muñoz-Piña et al. 2008, Pagiola 2008). The long-term viability of Mexico's payments for carbon sequestration services will depend upon international investors' willingness to support the communities currently receiving payments over 5 years for tree planting or forest conservation under the early normative frameworks (2004–2007) and those who have only been supported to design CDM-PINs with no current funding for implementation.

Service intermediaries should develop more effective marketing efforts to link national carbon forestry projects with the CDM and voluntary markets, which would in turn involve some re-adjustments in project design and implementation in order to adapt to the methodological standards governing these markets (Peskett et al. 2007). Even so, this may not be sufficient, as CDM investments in the forestry sector remain limited and voluntary offsetting markets are moving towards investments in renewable energy and energy efficiency measures (Corbera et al. 2009a). Alternatively – and following the Costa Rican example – the government could tax fuel-based power stations and provide Mexican companies with incentives to invest in PES projects, or develop a national emissions trading scheme encompassing offsetting through reforestation and forest conservation activities.

PES programmes draw important lessons for the future development of REDD activities in Mexico – particularly on the livelihood impacts of land-use change activities and the challenges involved in transferring conditional payments across actors and scales. According to CONAFOR, PES programmes will contribute to the sequestering of 1500–3100 million tCO_2e in standing forests up to 2012 and 0.08–0.16 million tCO_2e in reforested sites. Considering the broader *Pro-Árbol* policy framework, all forestry-related programmes may avoid the release of 11 000–21 000 million tCO_2e, and promote the sequestration of 18–42 million tCO_2e (Comisión Intersecretarial sobre Cambio Climático 2007). Furthermore, Mexico's future REDD programme can gain from the skills acquired at different governance levels in the design and development of PES programmes, including the identification of priority funding areas and the design of incentive-transfer mechanisms involving the federal government, social organizations, rural communities and local farmers in an even-handed way. In fact, CONAFOR is already considering whether REDD incentives could be used to extend the financing of those PES projects nearing the end of their 5-year implementation support period, especially those which are located in high deforestation risk areas.

The PES experience, however, shows that paying communities or farmers to enhance and adopt sustainable forest management practices can also exclude resource users who do not have formal property rights or who have limited power to shape decisions about the distribution of incentives, such as the case of women and informal right holders in *Niños Héroes*. The property rights underlying the stocks and flows of ecosystems and their services impact upon the design of PES schemes, including who gets involved, who holds the rights over the services and the payments, their transaction costs, all of which have welfare implications (Boyd et al. 2007). It cannot be assumed that PES or REDD incentives will equitably favour all community members, and careful attention needs to be given to decision-making processes across actors and scales.

It is worth noting a number of gaps and areas for future research. First, it is necessary to provide more consistent and structured data on PES impacts on each of the five 'capitals'. Fieldwork sampling and mapping techniques can be used to trace changes in the composition of ecosystems and the varying degrees of conservation over time in project locations. The impacts of PES on labour, time-allocation, and on households' education and skills endowments are under-researched, and they need to be studied and presented coherently across case studies and large-scale surveys for programme evaluations, using both quantitative and qualitative approaches. Information on investments in physical capital needs to be described more accurately, whilst data on financial capital should include indicators such as average income and transaction costs per household, financial distribution mechanisms and investment strategies, and a comparison of project design and implementation costs per hectare under different types of tenure regimes in order to differentiate between services and their associated costs. Research on social capital should investigate whether PES generates conflict among and between participants and non-participants, identify the root causes of such conflict beyond PES, and describe the level of trust among programme participants and how it changes over time. Furthermore, future research should describe how PES creates or shapes local institutions for natural resource management, and how these institutions stand the test of time.

Finally, understanding how PES incentives induce behavioural change at household and community levels needs to be further documented. The existence of single or combined incentives, such as financial benefits, social norms, compliance and cultural values may influence people's commitment to environmental conservation, and changes in the nature or the balance between different kinds of incentives for local resource users can also lead to behavioural changes, both collective and individual (White and Martin 2002; Gowdy 2008). Of course, such changes may also be driven by other variables, such as demographic pressures, changes in local leadership or migration trends, which will also shape the way people will manage and protect forests in the future.

ACKNOWLEDGEMENTS

This article would not have been possible without Lucía Madrid and Carmen González Soberanis, who developed most of the fieldwork and collected up-to-date information on PES programmes in Mexico. I am grateful to all individuals and organizations who participated in the interviews and the focus groups, particularly Mexico's INE and CONAFOR. The research has been developed with the financial support of the Australian National University and the International Development Programme of the Tyndall Centre for Climate Change Research. Any errors or omissions remain my sole responsibility.

NOTES

1. This chapter is an updated version of Corbera et al. (2009a).
2. Before the creation of these programmes, Mexico hosted one of the earliest carbon forestry projects in the world (Tipper 2002), and a watershed scheme in the municipality of Coatepec, which was one of the first in Latin America to establish a trust fund through which water consumers rewarded forest managers for the maintenance of forest cover upstream of the local hydrological basin (Manson 2004).
3. In *ejidos* and indigenous communities, household heads are entitled to a parcel of land which, originally, could only be bequeathed to a single descendant or spouse. Households share an area of communally owned forest and pasture over which a series of management regulations applies. Collective governance takes place through a collective assembly in which all rights holders participate and have a right to vote. The assembly then establishes the rules governing resource use and land distribution, among other issues.
4. The community assembly is the most important governance organization in Mexican *ejidos* and indigenous communities. The assembly periodically brings together all formal and often informal rights holders, who discuss and reach collective decisions concerning the community's life, including participation in development projects and programmes.
5. One tonne of carbon equivalent is a measurement unit equalling the concentration of carbon dioxide that would cause the same amount of temperature change in the climate system as the given mixture of carbon dioxide and other greenhouse gases.
6. One CER unit equates to one tonne of CO_2 being avoided or sequestered through a carbon project registered with the CDM (http://www.unfccc.int).
7. By beneficiary, the authors mean the *ejido*, association or private landowner which has received support from PSA-CABSA. In *ejidos* they conducted only a single survey, usually responded to by one member of the *ejido* authority.
8. A previous assessment – based only on 20 interviews with PSA-CABSA beneficiaries – notes that more than 80 per cent of participants earn less than US$190/month (Gómez Guerrero 2006, p. 75).

REFERENCES

Alix-Garcia, J., A. de Janvry, E. Sadoulet, J.M Torres Rojo, J. Braña Varela and M. Zorilla Ramos (2005), *An Assessment of Mexico's Payment for Environmental*

Services Program, Comparative Studies Service Agricultural and Development Economics Division, United Nations Food and Agriculture Organization (FAO), Rome: FAO.

Bezaury-Creel, J. (2007), El Consejo Técnico Consultivo (CTC) en el marco del Proyecto de Servicios Ambientales del Bosque, Payments for Ecosystem Services in Mexico: Current Status and Future Objectives Workshop, Mexico City: Tyndall Centre and National Institute of Ecology.

Boyd, E., E. Corbera Kjellén, M. Gutiérrez and M. Estrada (2008), The politics of 'sinks' and the CDM: a process tracing of the UNFCCC negotiations (pre-Kyoto to COP-9), *International Environmental Agreements: Politics, Law and Economics* 8, 95–112.

Boyd, E., P. May, M. Chang and F.C. Veiga (2007), Exploring socioeconomic impacts of forest based mitigation projects: Lessons from Brazil and Bolivia, *Environmental Science and Policy* **10**, 419–33.

Cibrián Tovar, J., V.E. Sosa Cedillo and L. Iglesias Gutiérrez (2008), Pro-Árbol Servicios Ambientales y Reducción de Emisiones por Deforestación y Degradación Forestal (REDD), Expo Forestal, Guadalajara, September 4–6, 2008.

Comisión Intersecretarial sobre Cambio Climático (2006), *Mexico: Tercera Comunicación Nacional sobre Cambio Climático*, Mexico City: Government of Mexico.

Comisión Intersecretarial de Cambio Climático (2007), *Estrategia Nacional de Cambio Climático*, Mexico City: Government of Mexico.

Corbera, E. (2005), Interrogating development in carbon forestry activities: a case study from Mexico. Doctoral dissertation, School of Development Studies, University of East Anglia, Norwich.

Corbera, E., K. Brown and W.N. Adger (2007), The equity and legitimacy of markets for ecosystem services, *Development and Change*, **38** (4), 587–613.

Corbera, E., M. Estrada and K. Brown (2009a), How do regulated and voluntary carbon-offset schemes compare?, *Journal of Integrative Environmental Research* **6** (1), 25–50.

de Vos, J. (2002), *Una tierra para sembrar sueños. Historia reciente de la Selva Lacandona 1950–2000*, Mexico City: Centro de Investigaciones y Estudios Superiores en Antropología Social y Fondo de Cultura Económica.

Ellis, F. (2000), *Rural Livelihoods and Diversity in Developing Countries*, Oxford: Oxford University Press.

Gómez Guerrero, A., A. Aldrete, A.M. Fierros González, G. Hernández Rivera, G. Ángeles Perez, J.I. Valdez Hernández, J.R. Valdez Lazalde, M.J. González Guillén, M.A. López López, P. Hernández de la Rosa, R. Rivera Vázquez and S. Fernández Cazares (2006), *Evaluación del Programa de Pago de Servicios Ambientales por Captura de Carbono, y los derivados de la Biodiversidad y para Fomentar el Establecimiento y Mejoramiento de Sistemas Agroforestales (PSA-CABSA). Ejercicio Fiscal 2005*, Guadalajara: Comisión Nacional Forestal.

Government of Mexico (2003), *Acuerdo Nacional para el Campo*, Mexico City: Government of Mexico.

Gowdy, J.M. (2008), Behavioural economics and climate change policy, *Journal of Economic Behavior & Organization*, **68**, 632–44.

Jaramillo, L. (2002), Estimación del Costo de Oportunidad del Uso de Suelo Forestal en *Ejidos* a Nivel Nacional, DGIPEA Working Paper, vol. 0205, México: Instituto Nacional de Ecología.

Klooster, D. (1999), Community-based forestry in Mexico: can it reverse processes of degradation?, *Land Degradation and Development* **10**, 365–81.

Klooster, D. (2003),'Campesinos and Mexican forest policy during the twentieth century', *Latin America Research Review*, **38** (2), 94–126.

Klooster, D. and S. Ambinakudige (2005), The global significance of community forestry in Mexico, in D. Barton Bray, L. Merino-Pérez and D. Barry (eds), *The Community-managed Forests of Mexico: The Struggle for Equity and Sustainability*, Austin, TX: University of Texas Press, 305–34.

Kosoy, N., E. Corbera and K. Brown (2009), Participation in payments for eco-system services: case studies from the Lacandon rainforest, Mexico, *Geoforum* **39** (6), 2073–83.

Manson, R.H. (2004), Los servicios hidrológicos y la conservación de los bosques de México, *Madera y Bosques* **10** (1), 3–20.

Martin, A., A. Blowers and J. Boersema (2008), Paying for environmental services: can we afford to lose a cultural basis for conservation? *Journal of Integrative Environmental Research* **5** (1), 1–5.

Martínez Tenorio, S., V. Sánchez Fabián and P.P. Ramírez Moreno (2007), *Evaluación externa de los Apoyos de Pago para Desarrollar el Mercado de Servicios Ambientales por Captura de Carbono y los Derivados de la Biodiversidad y para Fomentar el Establecimiento y Mejora de Sistemas Agroforestales (PSA-CABSA). Ejercicio 2006*, Guadalajara and Mexico City: Comisión Nacional Forestal y Universidad Autónoma Chapingo.

Millennium Ecosystem Assessment (2005), *Ecosystems and Human Well-being: Synthesis*, Washington DC: Island Press.

Muñoz-Piña, C., A. Guevara, J.M. Torres and J. Braña (2008), Paying for the hydrological services of Mexico's forests: analysis, negotiations and results, *Ecological Economics* **65**, 725–36.

O'Brien, K.L. (1998), *Sacrificing the Forest. Environmental and Social Struggles in Chiapas*, Boulder, Colorado: Westview Press.

Pagiola, S. (2008), Payments for environmental services in Costa Rica, *Ecological Economics* **65**, 712–24.

Patrinos, H.A. and E. Skoufias (2007), *Economic Opportunities for Indigenous Peoples in Latin America*, Conference Edition, Washington DC: The World Bank.

Peskett, L., C. Luttrell and M. Iwata (2007), *Can Standards for Voluntary Offsets Ensure Development Benefits?*, Forestry Briefing 13, London: Overseas Development Institute.

Rudel, T.K. (2008), Meta-analyses of case studies: a method for studying regional and global environmental change, *Global Environmental Change* **18**, 18–25.

Ruiz, F. (2007), Experiencias en la formulación de proyectos forestales de captura de carbono, Payments for Ecosystem Services in Mexico: Current Status and Future Objectives Workshop, Mexico City: Tyndall Centre and National Institute of Ecology.

Sen, A. (1999), *Development as Freedom*, Oxford: Oxford University Press.

Tipper, R. (2002), Helping indigenous farmers to participate in the international market for carbon services: the case of Scolel Te, in S. Pagiola, J. Bishop and N. Landell-Mills (eds), *Selling Forest Environmental Services. Market-based Mechanisms for Conservation and Development*, London: Earthscan, pp. 223–34.

White, A. and A. Martin (2002), *Who Owns the World's Forests? Forest Tenure and Public Forests in Transition*, Washington DC: Forest Trends and Centre for International Environmental Law.

Wunder, S., S. Engel and S. Pagiola (2008), Taking stock: a comparative analysis of payments for environmental services programs in developed and developing countries, *Ecological Economics* 65, 834–52.

4. Diversifying livelihood systems, strengthening social networks and rewarding environmental stewardship among small-scale producers in the Brazilian Amazon: lessons from *Proambiente*

Wendy-Lin Bartels, Marianne Schmink,
Eduardo Amaral Borges, Adair Pereira Duarte
and Hilza Domingos Silva dos Santos Arcos

INTRODUCTION

The Brazilian Amazon, which comprises nine states and covers 5.3 million km^2 (Soares et al. 2006), is vulnerable to interactions between economic, ecological and climatological factors that could lead to a 31 per cent reduction of its closed-canopy forest by 2030 (Nepstad et al. 2008). Scenario models predict that the current rapid expansion of mechanized agricultural crops, such as soybeans, and future prospects for obtaining ethanol from sugar cane, will likely induce cattle ranchers to move further into forested areas, increasing pasture land and reducing canopy cover. A subsequent increase in the number of forest fragments and fire sources may then lead to greater emissions of atmospheric aerosols that, in turn, will inhibit rainfall. The outcome of these interactions is a large-scale forest dieback that would potentially release 15–26 petagrams[1] of carbon, which could be further exacerbated by extreme climatic events (Soares et al. 2006; Nepstad et al. 2007; Nepstad et al. 2008). Therefore, decisions made over the next few years regarding conservation and development in Brazil may be the last opportunity to avert the predicted advance of significant drying and deforestation that threatens resilience, biodiversity conservation and the provision of ecosystem services within the Amazon (Mahli et al. 2008).

Proambiente is a rural development programme for the Amazon that integrates conservation into the development agenda by re-valuing traditional rural livelihood systems and rewarding sustainable land stewardship. *Proambiente* is unique in addressing family production systems as a whole, and uses incentives and capacity-building tools to enhance planning and the implementation of sound management practices by smallholders.

Proambiente provides cash payments to farmers who develop long-term management plans for their properties that comply with environmental guidelines and sign collective agreements promising to abide by them. Within these management plans, participants incorporate sustainable practices intended to deliver a bundle of environmental services and processes (Gomes et al. 2008), including reduced deforestation, increased carbon sequestration, restoration of ecosystem hydrological functions, conservation of biodiversity, reduction of potential nutrient loss and a reduction in landscape flammability (MMA 2005). Although partially conceptualized as a payment for environmental services (PES) programme, *Proambiente* is designed to incorporate a suite of additional social benefits that go beyond mere financial compensation of participating families; the programme also promises integrated technical assistance and credit programmes for smallholders, and their insertion into local and regional development planning processes.

Currently, *Proambiente* is being implemented in 11 experimental sites across the Amazon and engages 4214 families in a process of environmental management and certification (Oliveira 2008). During 2006, a total of 1768 *Proambiente* families received financial payments (Oliveira 2008) for developing land management plans and initiating changes in their livelihood systems.

This chapter reviews the development and implementation of *Proambiente* within the broader historical context of conservation and development in Brazil. We present preliminary impacts on livelihoods and governance structures in the state of Acre, as well as challenges and opportunities that the programme presents for a more sustainable Amazon region. Although we refer to general characteristics of *Proambiente* as they apply to all participating states, we specifically consider impacts within the state of Acre. The data presented in this chapter draw from research on *Proambiente* across the Amazon and, in particular, from dissertation fieldwork conducted by one of the authors, Bartels, who began researching the programme in 2004. In partnership with the local *Proambiente* implementing agency in the state of Acre (PESACRE), the researcher conducted 76 interviews between January and August of 2006 with participating *Proambiente* families in the communities of Nari Bella Flor and Porto Carlos (n = 47), extension agents (n = 6), and representatives

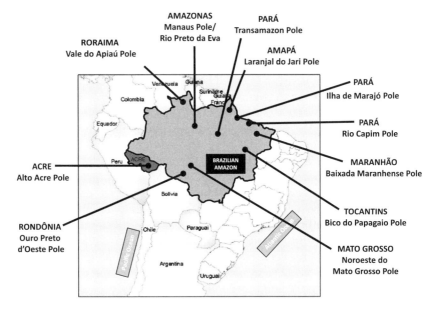

Source: Map adapted from Souza et al. 2006.

Figure 4.1   Map of Amazonian states, indicating the 11 original poles of
Proambiente

from government institutions and non-government organizations (NGOs) involved in the programme at the state level (n = 23).

Although the establishment of a legal framework for PES payments and a sustainable funding system for *Proambiente* remain problematic, it has been argued that the programme is a strong candidate for international funds from reduced deforestation and degradation (REDD) (Hall 2008) and could potentially introduce local communities to the opportunities that can be derived from directly linking with global initiatives focused on sustainable development and climate change mitigation.

THE DESIGN OF *PROAMBIENTE*

The initial *Proambiente* proposal included a plan for experimental sites (referred to as 'poles'), located in nine Amazonian states. However, in many poles, implementation has been stalled. Figure 4.1 indicates the original 11 poles that began the programme in 2004. In Acre, where this research was carried out, favourable state-level policies have supported continued *Proambiente* implementation. Indeed, *Proambiente* families are now included in a new state level programme that certifies sustainable production systems and the restoration of degraded land.

A total of 4214 families are registered with *Proambiente*. Family selection was not focused primarily on environmental criteria. In fact, *Proambiente* families were targeted mostly in accordance with their social and political associations (Araujo 2007; Hall 2008), since social organizations[2] carried out the selection. Families in the same geographical location or 'community' were organized into groups. Sixteen community groups were formed in the Acre pole, which are located within four municipalities in the Alto Acre region. Within community groups, selected families do not necessarily live adjacent to one another, but can be spatially fragmented by non-participating small farmers or large ranchers. Some households are near highways and small towns, with convenient access to markets, whereas others are extremely isolated and accessible only by boat.

Participating households have diverse family production systems including agroforestry, cultivation of subsistence crops (for example, beans, rice and manioc), market-oriented cattle-raising and/or forest extractive activities such as Brazil nut gathering, fruit collection, rubber tapping or fishing. Private land tenure is not a pre-requisite for participation in *Proambiente*. In Acre, 60 per cent of *Proambiente* households live in extractive reserves, which are located on state land with 20–30 year concessions granted to a community. Other *Proambiente* families live in colonist areas created by agrarian reform programmes for migrants, and practise small-scale farming. These families are allocated full private property rights which can be sold. One community group of *Proambiente* families in Acre resides neither in colonist areas nor reserves, but instead awaits official recognition of title to land that they independently settled and still occupy.

Stakeholder Roles and Responsibilities

Families participating in *Proambiente* are 'ecosystem service providers/ sellers', whereas the Brazilian government could be considered the current 'buyer'. Intermediary stakeholders include local community agents, teams of technical extension agents, state-level management councils, the Ministry of Environment (as the federal coordinating agency), and a national level Management Council. Extension agents at the state, local and community levels coordinate activities. The complex set of interacting organizations is shown in Figure 4.2. These include local level facilitators (that is, within communities), municipal-level extension agents, and state-level additional technical extension agents who guide implementation activities, source funding, etc. In Acre, the implementing agency is an

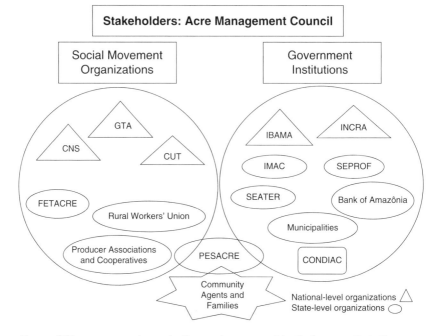

Notes: * Non-government organizations and government institutions constitute the management council (participating families are represented on the council by their associations and unions).
NB: Government agencies: IBAMA – Brazilian Ministry of the Environment's Enforcement Agency; INCRA – Brazil's National Institute for Colonization and Agrarian Reform; IMAC – State of Acre Secretary of the Institute of the Environment and Natural Resources; SEPROF – State of Acre Secretary of Family Production and Extractivism; SEATER – State of Acre Secretary of Technical Assistance and Rural Extension; CONDIAC – Consortium of Inter-municipal Development of the Upper Acre River and Capixaba.
Non-government organizations: GTA – Brazil Amazon Working Group; CNS – National Rubber-tappers Council; CUT – Unique Workers' Center (the chief union federation in Brazil); FETACRE – Federation of Rural Workers in Acre; PESACRE – Group of Research and Extension in Agroforestry Systems of Acre.

Figure 4.2   Proambiente *stakeholders in Acre**

NGO named PESACRE, which also arranges meetings of the state-level Management Council.

A federal level Management Council was formed in 2002 with the goal of consolidating the programme at the federal level, particularly in the political sphere. As with state-level institutions, this national council comprises several government and civil society organizations. *Proambiente* is housed within the Ministry of Environment, where a federal management team administers the programme and develops operational norms and regulations.

The Implementation Process

Proambiente was designed to be implemented in six phases over 15 years, starting in 2003. Each phase encompasses participatory planning and utilizes specific planning tools (see below) that engage a variety of stakeholders in future-oriented visioning processes.

Phase 1: Creation of a regional sustainable development plan to be implemented through the state management council.
Phase 2: Development of management plans for each family, created at the property level by household members, community agents and technical extension agents.
Phase 3: Implementation of the plans referred to in Phase 1 and 2, guided by rural technical extension agents.
Phase 4: Establishment of community agreements and certification plans.
Phase 5: Audits and certification.
Phase 6: Remuneration of environmental services by society.

Since 2003, when the programme began its official implementation, poles have advanced at different rates. By 2006, six poles had reached Phase 4 and established community agreements in 108 communities to engage a total of 2084 households (Oliveira 2008). However, some poles were still in Phase 2, with technical agents collecting data for developing individual management plans.

An important step in the planning process is the community agreement signed by a group of families who agree to be collectively responsible for negotiating the particular management practices that they will follow. Technical extension agents work with community agents and families to identify these specific management practices that will enable families to achieve their economic aspirations while being environmentally sustainable. (Ideally, the activities stipulated in these plans should shape technical extension services and link families with appropriate state-wide policies.) Signing the community agreement commits families to implement their property management plans, which include maps of current land uses and show both their proposed changes in the production system and their conservation goals.

Proambiente intends to implement a system of 'socio-environmental' best practice certification in two steps. The first 'participatory certification' step is shaped by the community agreements in which families monitor one another according to the requirements of the community agreement. In the second step, an independent certifier will be contracted to conduct audits of the transformation of conventional production systems into the sustainably managed agro-ecological system described in the property management plans (Oliveira 2008). The Brazilian standard setting agency, INMETRO (*Instituto Nacional de Metrologia, Normalização e Qualidade Industria*) is expected to assist in the design of certification standards for *Proambiente*, while an NGO, possibly IMAFLORA (*Instituto de Manejo e Certificação Florestal e Agrícola*), will conduct the audits. A timeline of key events in the Acre pole is represented in Figure 4.3. By 2008 (5 years after the initial engagement of families), the Acre pole had established community agreements in all 16 groups and was initiating the certification phase of the programme. During 2006, even before they reached the remuneration phase of the programme (Phase 6), families received payments referred to as 'help for costs' that were intended to reward them for initiating changes in their livelihood systems.

Monitoring of the Environmental Services

Monitoring the provision of environmental services and of adherence to community agreements was to be carried out by 'municipal commissions' comprising local social movement leaders and government institutions. However, indicators were not developed and there is currently no evidence that any monitoring has taken place in Acre.

A clearly formulated sliding scale for sanctioning non-compliance was established in project documents. For example, if a family joined a community agreement but failed to comply with the requirements for the first audit, the family would only receive 75 per cent of the payment. If, by the next audit, the family still had not changed practices, they would receive 50 per cent and so on, until, by the fourth time, the family would be eliminated from the programme. Such an approach to administering sanctions recognizes that farmers need time to adjust to the new management practices (Mattos 2004). In reality, however, conditionality within *Proambiente* is weak and families received initial payments regardless of whether or not they followed their community agreements (Mattos et al. 2006).

PES Payments: the Nature of Rewards

The original *Proambiente* proposal promised to pay families R$150 per month (US$71[3]) for the provision of environmental services, approximately

Preparatory Phase

2000
- *Grito da Amazônia:* The Amazonian social movement presents the *Proambiente* proposal to the Federal Government.

2003
- *Proambiente* is incorporated into the Federal Government's 4-year national development plan: *Plano Plurianual* (PPA) 2004/2007.
- In Acre, PESACRE, the implementing agency is contracted by the social movement (FETACRE).
- Development of a diagnostic instrument of the Acre pole, forerunner of a regional development plan.

2004
- Selection and registration of the *Proambiente* families and community agents.
- Formation of the state management council.
- Development and digitalization of the family and community diagnostic tools.
- Distribution of programme educational materials to stakeholders.

Negotiation Phase

2005
- Development and digitalization of individualized management plans for 400 families.
- Creation of community agreements.
- Creation of the Acre Council for Territorial Development (housed within *Proambiente*'s management council).

Implementation Phase

2006
- Families in Acre receive the first payment 'Help for costs' – for implementating management plans.
- State-wide assembly of the 400 Acre families held in the municipality of Brasiléia.
- Technical assistance with families to diversify production.
- Revision of individual management plans.
- Idea exchanges among participating families and community groups.

2007
- *Grito da Terra:* Social movement proposal to create a fund within the Federal Government's *Plano Plurianual* (PPA) 2008–2011 to maintain the activities of Proambiente.
- In Acre, 400 participating farmers plant green manures and invest in agroforestry systems.

2008
- Working group led by the Ministry of Environment drafts a legal framework to pay for environmental services.
- Creation of the Amazon Fund.
- State of Acre passes a new policy entitled 'Valuing the Active Forest Environment' inspired by *Proambiente*.

Figure 4.3   The evolution of Proambiente *in the Acre Pole 2000–08*

half of the average Brazilian minimum monthly wage in 2005. However, a principle bottleneck of *Proambiente* is the absence, not only of a legal and juridical framework to pay for environmental services in Brazil, but also the difficulty in financing such payments. By the end of 2007, six different proposals had been drafted and submitted to the National Congress in relation to establishing a legal PES framework in Brazil. Until the legal framework is established, money cannot be allocated from the public budget to pay farmers. To side-step these issues, the programme orchestrated a payment termed a 'help for costs'. Instead of paying for the provision of environmental services, farmers were rewarded for implementing their management plans and initiating land-use changes. These payments were intended to demonstrate the government's commitment to *Proambiente* and to act as an incentive for farmers to remain engaged until a legal mechanism is established.

In 2006, the state of Acre paid participating *Proambiente* families R$600 (US$282) in two instalments of R$300 (US$141) each, a rate of R$100 (US$47) per month. The Acre pole was unable to continue payments because a sustainable mechanism for generating these funds had not been developed at the federal level. Oliveira (2008) indicates that 1768 families from five *Proambiente* poles received cash payments during 2006. During this time, the federal government spent R$1 825 663 (US$861 163) in these poles, an average of R$1033 (US$487) per family. However, this estimate appears to include expenditures on meetings and other project implementation expenses and should not be assumed to constitute actual cash payments to farmers.

Importantly, financial payments are only one of the rewards that the programme promises. Participating families were also assured of in-kind, non-monetary advantages, for instance, technical assistance for livelihood system diversification and the development of long-term land-use plans, strengthened connections between participating families and local government to support families in securing land tenure, accessing specialized credit, and improving local infrastructure (for example, road building or grading, as well as health and education facilities).

Transaction Costs

Currently, *Proambiente* lacks institutional arrangements that provide reliable financial support to cover transaction costs. In the original proposal, several sources of funding were anticipated to cover programme costs – an environmental fund was to be set up to pay for environmental services, credit was to be provided to families by the government-subsidized credit fund, the *Fundo Constitucional de Financiamento do Norte*, and the

Table 4.1   Costs associated with Proambiente *since 2005 in the State of Acre*

Period Planned	Period executed	Federal funds (R$)	Source of resources	PESACRE funds (R$)	Total value (R$)
12/2007– 11/2008	12/2007– 05/2009	162 493	MDA/SAF	935	163 374
02/2006– 01/2007	02/2006– 05/2007	300 000	MDA/SAF	34 600	334 600
12/2005– 07/2006	12/2005– 07/2006	348 400	MMA/ Agroext.	45 000	393 400
08/2005– 12/2005	08/2005– 12/2005	213 000	MMA/ FNMA	0	213 000
02/2005– 07/2005	02/2005– 07/2005	180 000	MMA/ FNMA	0	180 000
Total		1 203 893		80 535	1 284 374

Notes: MDA/SAF – Ministry of Agrarian Development, Secretary of Family Agriculture; MMA/Agroext. – Ministry of Environment, Agro-extractivism programme; MMA/FNMA – Ministry of Environment, National Environmental Fund.

Ministry of Environment was to cover the costs of extension services from a second fund.

In practice, securing sources of funding has been complicated. When *Proambiente* began, finances came primarily through foreign dona-tions and several government departments (the Ministry of Social Development, the Ministry of Environment and the Ministry of Agrarian Development). Currently, implementing agencies in each pole are respon-sible for developing and presenting proposals to the national govern-ment to finance particular phases and activities, and agencies must then await federal government approval before commencing implementation. The extensive government bureaucracy in Brazil can result in a time-lag between when money is allocated at the federal level and when it is released at the state level. During these intervals, or when federal funds have been insufficient, PESACRE has relied on international funds to cover costs (USAID and an Italian NGO, Intervida). The Acre pole has also had recent financial support from the state government (PESACRE, pers. comm. 2008).

Table 4.1 indicates the costs associated with the programme since 2005. Between February 2005 and April 2009, R$1 284 428 (US$588 512[4]) was spent in Acre to implement *Proambiente*, delivered through the federal government and supplemented with funds leveraged by PESACRE.

Expenditures were used for rural extension to families, the 'help for cost' payments to families to encourage them to initiate sustainable practices on the properties, infrastructure development (such as the construction of a manioc production house), wages for technical agents, the development of individualized management plans for each family, and the development of diagnostic surveys of families and communities.

Working with a large number of small-scale farmers and members of forest-dwelling communities scattered across the Amazonian landscape leads to high transaction costs. These costs could be reduced if the programme engaged fewer farmers and those with larger properties. However, the goals of the programme are not solely based on ecosystem services or environmental conservation, but rather on promoting social justice. Therefore inherent to the programme's design is a tolerance of high transaction costs in order to serve marginalized small-scale resource users (Hall 2008).

The unstable financing has generated frustration among local staff and disillusion among farmers. Members of PESACRE and the state Management Council have raised concerns about the danger of increasing expectations among families, community agents and technical staff, and the potential loss of credibility in *Proambiente* when these hopes are not met. For instance, the inconsistent payment of salaries to technical agents has resulted in the loss of well-trained extension agents in Acre, who have been forced to look elsewhere for employment. In addition, when their salaries are interrupted, community agents discontinue visiting other *Proambiente* families within their community groups, limiting the scope of the programme's influence and resulting in the emergence of doubts among families regarding the programme's continued implementation.

THE IMPACTS OF *PROAMBIENTE*

In the following analysis, the preliminary qualitative impacts of the *Proambiente* programme are presented, based on interview data from 2006 and on recent consultations with key informants about emerging results at the household and state levels in Acre.

Impacts on the Environment

Extension agents encouraged farmers to invest their 'help for costs' payments (received in 2006) primarily in reforesting riparian zones, household orchards and agroforestry systems, practices being promoted by extensionists as ways to diversify livelihoods and stabilize deforestation among small-scale farmers. *Proambiente* families in Acre were encouraged to plant fast-growing native species with nutritional and economic values in an effort to integrate long-term environmental benefits (recuperation of degraded lands and carbon sequestration) with short-term social and economic yields that accrue to the family sooner. To facilitate these reforestation efforts, PESACRE also worked in each *Proambiente* community group to build greenhouses where families raised native seedlings for distribution in deforested riparian zones and on degraded lands. Through a partnership with the Acre Forest Secretary, PESACRE distributed 6000 forest tree seedlings to *Proambiente* families. By effectively harnessing this local government partnership, *Proambiente* did not need to fund the entire reforestation project.

Local investments in orchards, tree plantations, improved forage grass and timber management in the Amazon are threatened by fire. Therefore, in *Proambiente*, much attention is placed on promoting controlled-burn techniques and reducing the use of fire for shifting agriculture. Recent progress in combating fire in Acre is one of the programme's most prominent contributions to environmental conservation.

Community agreements among *Proambiente* families have set these 400 households apart from other families in the state. In 2005, when these agreements were developed, families negotiated new fire management practices. During that year, most families in the Alto Acre region were particularly sensitive to the issue of fire because of a severe drought, which was followed by the spread of uncontrolled fires that left 300 000 ha of primary forest in the south-western region of the Amazon damaged and more susceptible to repeated burning. The direct economic losses caused by these fires was estimated to be at least US$50 million (Brown et al. 2006). The damage caused by the fires compounded water scarcities associated with the drought and subsequent losses in their productive systems (for example, Brazil nuts, rubber, annual crops and animals) (PESACRE 2006). Thus, many participating *Proambiente* families that were victims of the fires became particularly receptive to *Proambiente* activities. Following this crisis, *Proambiente*'s Management Council capitalized on partnerships with the state university, the local environmental agency, and the Foundation for Technology Development in Acre to prioritize *Proambiente* community agents and families, linking them to municipal-level capacity-building efforts for fire control and monitoring. Once again, this state-level network demonstrated its potential to benefit families directly with promising conservation outcomes.

In addition to training in controlled-burn techniques, *Proambiente* has offered technical support for planting leguminous green manures to substitute the need for fire in shifting cultivation. Green manure systems can ensure soil fertility by adding organic matter and nutrients to the soil, especially when practised in rotation. In *Proambiente*, such systems were implemented on abandoned but previously cultivated land, so that farmers would not need to further deforest their legal reserve (a tangible environmental outcome). During 2007, 400 ha (1 ha per *Proambiente* family) were sown with green manures.

In 2008, the *Ministério Público* (Public Defender) of Acre passed an ordinance forbidding the licensing of any fires because of concern about 'hot spots' or areas particularly susceptible to burning following the 2005 drought. *Proambiente* had given participating families an advantage through their establishment of green manures in 2007, prompting the state government of Acre to become interested in expanding these sustainable practices to other communities. (The example reveals that *Proambiente* can potentially affect a broader segment of society and not only the families that are currently selected.)

Despite preliminary successes in training producers in fire-management techniques, encouraging reforestation of riparian areas and planting green manures, the programme is unable to guarantee the provision of environmental services. Because participating families were not selected in locations adjacent to one another and are sometimes separated by large ranches, fire originating from non-participants may escape onto *Proambiente* properties. This could threaten the forest reserve of participating families and possibly diminish any contributions they may have made within the programme. A suitable solution has not yet been devised to prevent participating families from being punished for the damaging activities of neighbours who do not participate in the programme.

Another concern is *Proambiente*'s lack of performance measurement and monitoring – baselines have not been set, a methodology for calculating the amount of carbon sequestered has not been determined and indicators have not been developed for evaluating changes in forested area over time. Changes in land management practices are not effectively supervised, with community members expected to voluntarily monitor one another. Considering the strong kinship ties in certain communities, it is unlikely that family members would denounce one another for utilizing fire or for failing to incorporate specific practices. This will need to be managed with institutional support from outside the community.

Within such a weak monitoring context, it is challenging for *Proambiente* to demonstrate that land management improvements are being made in areas that are largely deforested. Without measuring environmental impacts, monitoring behaviour change, or developing conditionality around payments, no direct link exists between payments, sustainable management and conservation outcomes.

Impacts on the Livelihoods of Participants

Financial capital

In 2006, participating *Proambiente* families in Acre received two payments of R$300 (total = R$600, US$282) to implement their management plans, as discussed above. In this way, *Proambiente* increased the average annual income of producers by around 14 per cent.[5] Unfortunately, these payments were not sustained beyond 2006, and because of this they do not represent significant welfare gains. However, the programme continues to provide technical assistance, encouraging the diversification of production systems. In addition to the establishment of additional sources of income, this approach encompasses household nutrition and wellbeing, not just income generation. In the long term, this may help to protect families from losses due to unforeseen market crashes of particular commodities or forest products.

Depending on their access to markets and potential for production, *Proambiente* families are integrated into state projects and emerging specialized markets in Acre that promise greater returns on products than conventional markets. For instance, in 2007, 70 *Proambiente* families were linked to the organic Brazil nut market; and 180 *Proambiente* families sold rubber to a new condom factory in the town of Xapuri, which paid producers 30–40 per cent more than could be gained from a conventionally sold block of rubber. Community-managed timber projects also enable *Proambiente* families to access markets for certified wood.

Proambiente has failed to meet expectations relating to access to credit for small producers. Though the programme originated as a response to failed loan schemes along the agricultural frontier in the Amazon, Brazilian lending agencies remain entrenched in financing traditional crops or agro-livestock systems that provide predictable income flows. Investing in smallholder agroforestry is perceived as too risky because of its long-term time horizon and the unpredictability of income yield. Despite the provision of eco-friendly federal credit, the Bank of Amazônia still requires extensive feasibility studies (Hall 2008). Ideally, banks would accept the management plans developed by *Proambiente* families as a part of the requirements for them to qualify for loans. Greater learning and flexibility is necessary beyond the property level, within banks themselves, before diversified low-interest long-term loans become a reality.

Human and natural capital

Stakeholders are investing time to learn and experiment with the new sustainably managed agro-ecological practices encouraged by *Proambiente*. Extension agents are gaining new skills, such as the ability to view landholdings holistically within a matrix of natural land; not merely focusing on individual crops. *Proambiente* has reshaped training for extension agents, moving away from the transfer of 'one-size-fits-all' packages for conventional high-input farming commodities and links to financiers. *Proambiente* has also included community agents (themselves farmers) in much of the training offered to extension agents, building their capacity and strengthening community social networks. The training provided has ranged from focusing on technical aspects of improved land management, participatory planning and social organization (for example, understanding dynamic group processes and participation within communities).

Human capital is also built by improving the collaborative decision-making capacity and skills of participating families, technical extension agents and members of the state Management Council. In addition to technical skills for adapting management practices, emphasis is placed on community organization and planning.

Although it is too early to determine the extent to which programme capacity-building efforts have influenced motivations for environmental protection, a prioritization exercise was conducted where families ranked *Proambiente*'s proposed benefits,[6] the results of which were (in order): environmental conservation, diversified technical assistance to improve household production, PES payments, access to credit and the participatory planning process. The fact that the PES payment was selected as the third most important could indicate that farmers may not view this sum of money as large enough to act as a primary incentive for behavioural change. Rather, participants consider rural extension efforts focusing on diversifying production and connecting farmers to specialized markets are more important contributions to livelihood change.

The specific barriers preventing farmers from adopting sustainable management practices are said to be a lack of finance (47 per cent), a lack of labour (34 per cent), lack of infrastructure (10 per cent) and a lack of knowledge or unfamiliarity with such sustainable practices (8 per cent). When asked what they would do with additional money, people mentioned hiring workers to cut green manures or buying fencing materials to cordon off riparian areas. Others pointed to a need for credit from the bank for more long-term investments. These results suggest that, in order to achieve sustainable land management practices, access to finance must accompany capacity-building efforts.

Political and social capital

The establishment of *Proambiente* has made a contribution to strengthening Brazilian democracy and building political capital. By bringing together a diverse array of stakeholders and successfully influencing public policy at the national level, an organized segment of civil society demonstrated their skills and commitment to influencing policy-making, even in the presence of the Amazon region's weak federal institutions (Mattos 2006).

At the micro level, management plans and community agreements foster dialogue between and among families and strengthen their organizational capacity and linkages to institutional networks. *Proambiente* families have gained access to state-government projects with the influence of representatives on the *Proambiente* Management Council and the provision of information by extension agents. These activities not only increase income generation, but also build relationships and strengthen network ties.

At the state level, the *Proambiente* Management Council in Acre has been instrumental in strengthening networks and coordinating actions among government institutions and NGOs. This dialogue has had unanticipated outcomes beyond *Proambiente* activities. For example, the Governor of Acre introduced a new policy in 2008, formulated after consultation among *Proambiente* Council members and a broader working group that met for regional ecological and economic planning and zoning. Largely inspired by the *Proambiente* experience, the new policy, 'Valuing forest environmental assets', proposes to certify 4000 sustainable rural properties and recuperate degraded areas in Acre. While certification is open to all families, *Proambiente* participants have the advantage of already being at the fourth (and final) stage in a certification process. The policy may also offer a state-level alternative for remunerating *Proambiente* families, but perhaps more importantly, the policy is a measure of successful collaboration at the state level and demonstrates the capacity of *Proambiente* leaders to influence policy-making.

The Livelihoods of Non-participants

Some families who are not included in a *Proambiente* group in Acre have expressed concern that, despite the fact that they follow the same sustainable management practices as many *Proambiente* families, their products will not be recognized in specialized markets because they will not have *Proambiente* certification. They also raise questions about how some families were selected, and why they were not. The response from extension agents is that the 400 *Proambiente* families are being rewarded for the entire *process* of certification itself, not only for their management practices. This process requires a substantial investment in time and a level of commitment through all programme phases.

Since the beginning of 2009, non-participating families in Acre find themselves situated within an enhanced policy environment that was inspired directly by the *Proambiente* experience. The recently approved state policy for certifying farms will open doors to 3600 non-*Proambiente* families.

External Influences and Future Research Needs

Proambiente has focused on integrating its goals with other state and local policies and projects. Thus some impacts are the direct result of these new connections, partnerships and institutional arrangements and it is difficult to disentangle direct causes and effects of *Proambiente*. Therefore, quantitative research is required to evaluate the specific ways that the programme affects its diverse stakeholder groups and the extent to which new institutional partnerships have proved advantageous to *Proambiente* families. It is also important to evaluate how the programme affects non-participating families at both the household and community level.

The programme lacks a methodology for measuring environmental services and for monitoring their continued provision. Suitable indicators must be identified to measure environmental change within *Proambiente* properties and across communities, and the programme must develop a system of accountability that links payments (or other programme benefits) with compliance to programme rules. Another aspect that deserves attention is the calculation of programme transaction costs, including the costs of developing and implementing monitoring and certification systems. Finally, more studies are needed to estimate the opportunity costs within the programme.

Research in the Amazon about estimating PES and REDD opportunity costs is in a formative stage. Opportunity costs have been calculated using spatially explicit rent models for high carbon retention (that is, selective timber harvesting) and low carbon retention (that is, soy and cattle ranching) uses of Brazilian Amazon forests and a cost of US$5.50 per ton of carbon has been predicted (Nepstad et al. 2007). This price can be reduced to US$2.80 per ton of carbon if the land where soy and cattle are highly profitable is omitted from calculations (6 per cent of the land area under consideration). A total of US$8 billion over 30 years would be sufficient to reduce carbon emissions from deforestation by 6 billion tons below the historical baseline in the Amazon (set at $20000\,km^2$ of deforestation per year) (Nepstad et al. 2007).

In this model, costs are based on private Amazon land development for crops and pastures, and the outcome depends heavily on distance to roads and on suitable soil and climate conditions (Borner and Wunder 2008). Therefore, opportunity costs are spatially disaggregated, as farms have different environmental and economic constraints. Borner and Wunder

(2008) show that payment modalities also affect cost estimations. For instance, if individuals are compensated for their opportunity costs using a maximum payment needed to avoid deforestation, CO_2 is predicted to cost US$13 per ton. However, when using the fluctuating prices of the Chicago Exchange, CO_2 would cost US$3.88 per ton and, in the most realistic scenario, one of temporary uniform fixed carbon credits, a ton of CO_2 would cost US$2.32. (Borner and Wunder 2008).

Opportunity costs in Acre may be affected by the new inter-oceanic highway that is currently under construction in this south-western Amazon region. This large infrastructure project has the potential to reshape economic, social and political conditions. On the one hand, increased transport could allow access to new markets for 'green' products and present viable alternatives for family production. On the other hand, the highway may affect land prices, stimulate speculation and cause an influx of new landowners with production practices that could increase forest fragmentation (Perz et al. 2008). Research is needed on the potential influences of the highway and specifically how it might alter opportunity costs in the region. *Proambiente* families will have to navigate new trade-offs that may ultimately affect the programme's ability to act as an incentive for sustainable land use.

CONCLUSIONS

Although *Proambiente* falls short of meeting the five criteria for a PES programme as defined by Wunder (2005, 2007), it has helped to establish the groundwork for the development of true PES programmes in Brazil. *Proambiente* suffers from many of the shortcomings that are sure to arise during the nascent stage of such schemes, especially under wavering political support, but to dismiss its impacts (at least in some pilot areas) would be to ignore its contribution to the advancement of the PES debate in the Amazon – *Proambiente* initiated discussions about paying for environmental services in Brazil and framed the issue as a national imperative. By increasing the salience of this issue at the national and state levels, *Proambiente* has paved the way for more recent programmes such as *Bolsa Floresta* in Amazonas, certification of farms in Acre and other proposals that support financial compensation of rural land owners, such as REDD.

While emphasizing *Proambiente*'s contribution to the debate on environmental services in Brazil, it is important to recognize the programme's weaknesses as a PES scheme. For example, the programme fails to isolate individual environmental services clearly, but rather rewards the implementation of land-use practices that are assumed to provide a bundle of environmental services. However, neither service provision nor land-use change is monitored effectively and *Proambiente* is unable to guarantee the continued provision of environmental services. In addition, families were rewarded without verification of compliance, indicating a lack of conditionality. Furthermore, the programme lacks a thoroughly planned auditing process for its certification process. The principal factor underlying these weaknesses is the lack of federal government support, which is demonstrated by an unreliable allocation of funds for programme activities. Without securing guaranteed funds, the national government has become an impotent 'buyer' of environmental services.

However, *Proambiente* was never designed to be a purely market-driven scheme and it was never intended that cash payments would provide the main incentive for family participation, which is a critical distinction between this programme and other PES initiatives. From the outset, the Amazon social movement envisioned a programme that would work through partnerships to provide families with a suite of benefits. In fact, this study shows that farmers are more interested in receiving technical assistance to improve productivity than in isolated payments. Apart from the technical assistance provided to farmers, which has influenced the substitution of legumes for fire in the agricultural systems of Acre, many of the promised benefits of *Proambiente* are yet to be realized. For example, although families were initially told that the programme would facilitate a line of credit to support investments in agroforestry systems and sustainable practices, no such credit option has been made available. In this example, there was a failure to integrate the goals of *Proambiente* with the objectives of other government agencies.

Some of the problems facing *Proambiente* arise because the programme was conceptualized and implemented quite rapidly, during a window of political opportunity. The programme's transformation from a social movement proposal to a government programme occurred before several aspects of its design (for example, the legal framework, funding and monitoring methodologies) were thoroughly considered and consolidated. The programme effectively became an experiment in adaptive management, and it was expected that learning and modification would occur during implementation. However, many complications emerged, exacerbated by weak political support from the federal government – the government did not adequately finance programme activities, nor did it establish a fund to pay farmers for environmental services. Even when money was released at the federal level, Brazil's complex bureaucracy caused a delay in those finances reaching the field and the programme lost credibility among local agencies, which made it difficult to harness partnerships across state agencies effectively. In Acre, *Proambiente*'s continued implementation is partly due to the involvement of key leaders from within the social movement who now occupy positions in the state government and reinforce inter-agency networks.

Proambiente demonstrates some of the challenges faced by efforts that aim to enhance democracy by drawing on grassroots demands to develop and implement public policy. Although the creation of *Proambiente* by the social movement is evidence of strengthened democracy in environmental policy-making, the implementation has demonstrated the difficulty of harnessing the strengths of government institutions, civil society and research organizations. *Proambiente*'s transformation from civil society proposal to government programme has left social movement leaders struggling to settle on a new role in the implementation process. These leaders now find themselves limited by a lack of federal government support, and pressured by the unmet expectations of farmer organizations.

Preliminary results in the state of Acre indicate that *Proambiente* has begun to affect the livelihoods of small-scale rural households in Acre as they initiate green manure cultivation, practise controlled-burn techniques, plant orchards and plan for future land-use changes on their properties. Participating families are also now embedded in new institutional arrangements at the community, municipal and state levels, allowing them greater access to information about government programmes and projects. In addition, *Proambiente*'s participatory land-use planning methodology has shaped recent public policy in the state of Acre – for example, the new state policy, 'Valuing forest environmental assets', has already secured financing to extend support to *Proambiente* participants and other small-scale producers across the state of Acre. Participating producers will develop and implement land-use management plans and comply with green certification requirements. In return, they will receive technical assistance and a financial bonus for following sustainable land-use practices. PESACRE is now working with other agencies to design a system of monitoring that will ensure accountability and conditionality at the state level for the certification of small farms. With these monitoring systems in place, families will be more favourably positioned to provide the deliverables required by REDD schemes.

Recent developments at the national level also indicate advances that could favour the resumption of payments to *Proambiente* families. The Ministry of Environment is creating a legal framework for PES and the Brazilian government created the Amazon Fund in July 2008, which will accept donations for the prevention, monitoring and fighting of deforestation, as well as to foster the conservation and sustainable use of forests in the Amazon biome. The Norwegian government has already committed

US$1 billion to this fund. This raises much hope for the sustainable funding of programmes such as *Proambiente*.

Additional proposals for the region include a strategy whereby global funders co-finance and complement Brazilian command-and-control strategies such as environmental law enforcement (Borner and Wunder 2008). These payments would be contingent on performance measures, monitored by a credible entity. Because *Proambiente* shows little evidence of implementing a monitoring mechanism at the project level, the verification of leakage, additionality and permanence remain troublesome. Despite these uncertainties, Fearnside (2008) urges against waiting for climate impacts to become more apparent before acting to value forest maintenance and reward environmental services in Brazil.

Although PES and REDD initiatives may reduce deforestation and environmental degradation, financial compensation alone is not likely to guarantee a better quality of life for smallholders in the Amazon. Benefits may need to incorporate critical socioeconomic services such as health and education that are needed by land-user groups who live in the areas under question (Mahli et al. 2008). In addition, DiGiano (2006) cautions that although PES payments in *Proambiente* could effectively raise incomes, payments alone may have little positive impact on forest conservation unless they are linked to alternative natural resource management practices, such as agroforestry systems. Indeed, payments alone may have perverse effects by providing financial resources for the expansion of cattle production and pasture area, a dominant livelihood trend in the study region (Gomes 2004). Within this context, an integrated package of benefits may be more feasible. It is unlikely that REDD initiatives can afford to pay the high opportunity costs associated with Brazil's agricultural frontier, where powerful commercial agro-industrial sectors lobby for cattle ranching, soybean and biofuel production. These highly profitable land uses affect land prices and can provide opportunities for small-scale families to sell their properties or adopt land uses that require forest conversion. Under such conditions, small producers may demand larger financial incentives to guarantee the provision of environmental services and sustainable land management.

Proambiente provides an innovative PES design for incorporating financial payments into a suite of other benefits such as vital social inputs needed to maintain rural people on their properties. Integrated REDD packages could include the delivery of reliable technical assistance and the creation of alternative markets through a greater commitment to policies that favour small-scale producers and diverse forest-based livelihood systems at both the federal and state levels. Promoters of REDD schemes have been urged to bring together stakeholders whose livelihoods are based on forest land or products, and to consult with them through a participatory planning process that builds on social capital among active civil society organizations (Johns et al. 2008). Therefore, incipient REDD programmes in the Amazon will benefit from analyzing *Proambiente*'s lessons in its endeavour to support integrated participatory planning processes that extend from the farm to regional levels.

ACKNOWLEDGEMENTS

We acknowledge valuable input from specific key Brazilian stakeholders, including Luiz Rodriguez De Oliveira of the Ministry of Environment, Luciano Mattos of the State University of Campinas, and IPAM researchers Ricardo Mello and Marcos Rocha. Their contributions were essential during the original field research and for recent data on programme implementation and impacts. We appreciate the assistance of families and staff of the NGOs, PESACRE (in the state of Acre) and FVPP (in the state of Pará). Research fieldwork was supported by the World Agroforestry Center (ICRAF, Belém) through Roberto Porro, as well as The Tropical Conservation and Development Programme at the University of Florida. Special thanks to Matthew J. Palumbo for editing.

NOTES

1. One petagram is equal to one billion metric tonnes.
2. *Proambiente* originally emerged in the 1990s; the main protagonists involved in its development included the Federation of Workers of Amazonia (FETAGs), the National Movement of Fishermen (MONAPE), the Coordination of Indigenous Nations of Amazonia (COIAB), the Institute of Environmental Research in the Amazon (IPAM) and the Federation of Organizations for Social and Educational Assistance (FASE). These grassroots movements received support from the Workers Party whose candidate, Luiz Inácio (Lula) da Silva, was elected Brazilian president in 2002 (Hall 2008). The Workers Party government subsequently transformed the social movements' proposal into a federal government programme (Allegretti 2007) and officially adopted *Proambiente* as a national programme by adding it to the federal government's 4-year development plan (Plano Plurianual 2004–2007).
3. Currency exchange estimated from March 2006, the month in which the first payments were made in Acre. $US1= R$2.12 on 15 March. Result rounded up. http://finance.yahoo.com/currency-converter.
4. At an exchange rate of $US1 = R$2.18 (17 April 2009) http://finance.yahoo.com/currency-converter
5. The average minimum wage per month in Brazil until March 2006 was R$300. In April 2006 it was raised to R$350. Source: http://www.portalbrasil.net/salariominimo.htm#sileiro. Therefore, for the 12 months of 2006, R$4050 was the total yearly minimum wage. The cash payment of R$600 is 15 per cent of this annual rate.
6. The question asked was: These five cards show the various benefits that the programme offers. Which is the most important to you and your family?

REFERENCES

Allegretti, M.H. (2007), Movimentos sociais e politicas publicas na Amazônia, paper produced for the Centre of Latin American Studies, Gainsville: University of Florida.

Araújo, (2007), A participação dos agricultores na construção do *Proambiente*. Uma reflexão a partir do Pólo Transamazônica, Masters' Thesis, Belem: Universidade Federal do Pará Centro de Ciências Agrárias.

Borner, J. and S. Wunder (2008), Paying for avoided deforestation in the Brazilian Amazon: from cost assessment to scheme design, *International Forestry Review* **10** (3), 496–511.

Brown, I.F., W. Schroeder, A. Setzer, M. de. L. R. Maldonado, N. Pantoja, A. Duarte and J. Marengo (2006), Monitoring fires in southwestern Amazonia rain forests, *EOS American Geophysical Union* **87**, 253.

DiGiano, M. (2006), The potential impacts of environmental service payments on smallholder livelihood systems in Brazil's Western Amazon, Master's thesis, Gainsville: University of Florida.

Fearnside, P.M. (2008), Amazon forest maintenance as a source of environmental services, *Anais da Academia Brasileira de Ciências* **80** (1), 101–14.

Gomes, C.V.A. (2004), Cattle ranching expansion among rubber tapper communities in the Chico Mendes extractive reserve in the Southwestern Brazilian Amazonia', Report for World Wildlife Fund-WWF-Brazil, Rio Branco: WWF.

Gomes, C.V.A., W.L. Bartels, M.A. Schmink, A.P. Duarte and H.D.S.S. Arcos (2008), Planejando Futuros Sustentáveis para os Pequenos Produtores: Programa *Proambiente*, Pólo Alto Acre, in Nurit Bensusan, O Manejo da Paisagem e a Paisagem do Manejo, Brasília: IIEB.

Hall, A. (2008), Better RED than dead: paying the people for environmental services in Amazonia, *Philosophical Transactions of the Royal Society B*, **363**, 1925–32.

Johns, T., F. Merry, C. Stickler, D. Nepstad, N. Laporte and S. Goetz (2008), A three-fund approach to incorporating government, public and private forest stewards into a REDD funding mechanism, *International Forestry Review* **10** (3), 458–64.

Malhi, Y., J.T. Roberts, R.A. Betts, T.J. Killeen, W. Li and C.A. Nobre (2008), Amazon climate change, deforestation, and the fate of the Amazon', *Science* **319**, 169.

Mattos, L. (2006), Capital Social na Concepção de Políticas Públicas: A importância socioeconômica e ecológica dos sistemas agroflorestais frente aos mecanismos de desenvolvimento, paper presented at the VI Congresso Brasileiro de Sistemas Agroflorestais, Universidade Estadual do Norte Fluminense, Rio de Janeiro, October 2006.

Mattos, L. (2004), Superando a Dicotomia entre Produção Rural e Conservação Ambiental: relato de uma iniciativa dos movimentos sociais rurais da Amazônia, in B. Pokorny, C. Sabogal and F. Krämer (eds), *Florestas, Gestão e Desenvolvimento: Opções para a Amazônia*, Brasília: CIFOR.

Mattos, L., A. Cau and P. Moutinho (2006), *Effectiveness of the Clean Development Mechanism within the Context of Forest Activities in Brazil: A Critical Analysis*, Development Dividend Task Force, Ottawa: International Institute for Sustainable Development.

MMA (2005), Proambiente: Certificação de Serviços Ambientais do Proambiente, Brasília, Brasil: MMA.

Nepstad, D.C., B. Soares-Filho, F. Merry, P. Moutinho, M. Bowman, S. Schwartzman, O. Almeida and S. Rivero (2007), The costs and benefits of reducing carbon emissions from deforestation and forest degradation in the Brazilian Amazon, available at www.whrc.org/BaliReports/ (accessed 23 August 2008).

Nepstad, D.C., C.M. Stickler, B. Soares-Filho and F. Merry (2008), Interactions among Amazon land use, forests and climate: prospects for a near-term forest tipping point, *Philosophical Transactions of the Royal Society* **363**, 1737–46.

Oliveira, Luiz Rodrigues de (2008), Serviços Ambientais da agricultura familiar: Contribuições para o desenvolvimento sustentável da Amazônia, Masters' Thesis, Brasilia-DF: Faculdade de agronomia e medicina veterinária, Universidade da Brasília.

Perz, S., S. Brilhante, F. Brown, M. Caldas, S. Santos Ikeda, E. Mendoza, C. Overdevest, V. Reis, J.F. Reyes, D. Rojas, M. Schmink, C. Souza and R. Walker (2008), Road building, land use and climate change: prospects for environmental governance in the Amazon, *Philosophical Transactions of the Royal Society B*, **363**, 1889–95.

PESACRE (2006), *Relatorio do Conselho Gestor*, Rio Branco, Acre: Municipality of Brasileia.

Soares-Filho, B.S., D.C. Nepstad, L.M. Curran, G.C. Cerquiera, R.A. Garvcia, C.A. Ramos, E. Voll, A. McDonald, P. Lefebvre and P. Schleisinger (2006), Modelling conservation in the Amazon basin, *Nature* **440**, 520–23.

Souza, C. Jr., A. Veríssimo, A. da Silva Costa, S.R. Reis, C. Balieiro, J. Ribeiro (2006), Dinâmica do Desmatamento no Estado do Acre (1988–2004), Report for Instituto do Home me Meio Ambiente da Amazônia.

Wunder, S. (2007), The efficiency of payments for environmental services in tropical conservation, *Conservation Biology*, **21** (1), 48–58.

Wunder, S. (2005), Payments for environmental services: some nuts and bolts, CIFOR Occasional Paper No. 42, Bogor: CIFOR.

5. The livelihood impacts of incentive payments for watershed management in Cidanau watershed, West Java, Indonesia

Beria Leimona with Rachman Pasha and N.P. Rahadian

INTRODUCTION

Payment for environmental services (PES) is now quite a well recognized approach in Asia. Interest and investment from international donors has enabled the testing of different PES mechanisms over the last decade, particularly those focusing on watershed protection and carbon sequestration. With the exception of China and Vietnam, where the schemes are state-run, schemes in Asia are generally small-scale, community-level projects.

The case study presented in this chapter is located in Cidanau, Indonesia. The Cidanau watershed is one of the most important watersheds supplying the domestic and industrial water needs of Banten Province, Java Island, Indonesia. The watershed covers 22 260 ha located between two regencies: Serang and Padeglang, and their six sub-districts. The Cidanau watershed also has a special role in biodiversity protection. In the base of the bowl-shaped Cidanau watershed lies the Rawa Danau Reserve – a 4200 ha nature reserve, which contains the only remaining lowland swamp forest in Java and has 131 endemic species. The reserve is important in the hydrological process, too, as the reservoir for the Cidanau River and its tributaries, which then flow into the Sunda Strait.

The Cidanau project was initiated by a multi-stakeholder watershed forum – *Forum Komunikasi DAS Cidanau* (FKDC)[1] and facilitated by the *Rekonvasi Bhumi* and the Institute for Social and Economic Research, Education & Information – both Indonesian non-government organizations (NGOs). In the beginning, the aim of the PES scheme was to slow down the environmental degradation of the Rawa Danau Reserve and the watershed around it. The PES scheme in Cidanau officially started in

Table 5.1   The sample of FGD participants

Village	Participating households	Proportion of total participating households (%)	Non-participant households	Proportion of total non-participating households (%)	Total households in each village
Cikumbuen	32	100	30	18	203
Citaman	43	100	30	18	210
Kadu Agung	38	100	30	8	414
Total	113		90		

2004 when a state-owned water company – the *Krakatau Tirta Industri* – and the FKDC, representing the upstream farmers, signed a contract to conserve the watershed.

This chapter describes the process of initiating the PES scheme and its design, and reviews the impacts of the 5-year scheme on local livelihoods. We assessed these impacts through a series of focus group discussions with the participants and non-participants and interviews with implementing agencies.

Methods to Assess the Impacts of the PES Scheme

We collected qualitative data from three villages in the Cidanau Watershed (Citaman, Cikumbuen and Kadu Agung). In each village, we held two focus group discussions (FGDs) for participants and two FGDs for non-participants. All the PES participants joined the discussion and, for the non-participants, we contacted village leaders who organized available household representatives to join the FGDs. The non-participants comprised 30 households from each village. In total, the FGDs involved 113 participants and 90 non-participants (Table 5.1).

The facilitators guided the FGDs through a series of questions on the impact of PES by comparing three time-periods: 1998–2000 (2000 was a landmark year, remembered by communities because it marked the beginning of political reforms and an economic crisis), the years between 2000 and 2004, and after signing a PES contract (2005 to the present). The livelihood impacts were discussed in terms of the five asset types covered in the Sustainable Livelihood Framework: financial, human, social, physical and natural. For each asset category, we asked the participants as a group to identify relevant impacts (Table 5.2), and collectively to rank them according to their relative importance. For example, under financial assets,

Table 5.2　The livelihood issues discussed in focus groups

Capital	Type of information discussed
Financial	Sources of income over the three periods
Human	What (if any) capacity/skills/knowledge were gained through the scheme?
Social	What was the nature and degree of trust with other stakeholders during the three periods? What norms or standards of behaviour did the community set itself in connection with the scheme (e.g. sanctions, etc.)? What were the community's networks like during the three periods?
Natural	What benefits did they gain from the watershed and its protection?
Physical	Had any investments been made as a result of the scheme (e.g. infrastructure)?

groups listed all sources of income during each era. The ten most important sources were then ranked, and paper dots were used by the facilitators to describe the relative percentage that each income source contributed to overall household incomes. Some impacts, such as trust and social capital, required further discussion to clarify their meaning.

In addition to the FGDs, a 1-day workshop was held involving FKDC members, local government and the Krakatau Tirta Industry company. We followed this up with some informal interviews to clarify any conflicting or unclear data from the workshop. In analysing livelihood impacts, the data are limited to the results from the FGDs and stakeholder interviews, as there has been no detailed quantitative analysis, so far, of household-level livelihood impacts in Cidanau.

THE DESIGN OF THE PES SCHEME

The Environmental Problems in Cidanau

The Cidanau watershed has been experiencing rapid change in land cover for almost two decades as forest is converted for agriculture due to population increase and a high dependence on farming.[2] The number of people living and farming illegally in the upstream protected area increased from around 600 in the late 1990s to an estimated 1500 in 2007. This period has also seen the conversion of conservation forest to rice fields and other crops. In addition, the Rawa Danau Reserve has experienced intensive encroachment and associated decreases in flora and fauna diversity. In 2000, about 20 per cent of the Rawa Danau natural reserve area had been encroached upon (Darmawan et al. 2005).

As noted earlier, this conversion of forest to farming land combined with unsustainable farming practices degrade the environmental services (ES) provided by the Cidanau watershed. The Cidanau watershed is the only water supply for the Cilegon housing and industrial area as well as for approximately 100 industries that operate around it. The main problems experienced by the water consumers (the ES beneficiaries) of the Cidanau watershed are shortage of water in the dry season and water quality degradation due to pollution and high sedimentation (Adi 2003; Munawir and Vermeulen 2007; Budhi et al. 2008).

Fluctuating water flow and water quality are the most important problems in Cidanau. The average discharge is 12.5 m³/s, fluctuating from an annual minimum of 1.2 m³/s in the dry season (August) to an annual maximum of 44 m³/s in the rainy season (January) (Adi 2003). In addition to the fluctuating water flow problem, intensive use of fertilizer and agricultural chemicals and the process of burning paddy husk reduce the quality of Cidanau's water. Remote sensing observation indicates that about 71 per cent of the watershed is prone to degradation, with the rate of erosion above 35 tonnes/hectare/year (Adi 2003). The sedimentation narrows water channels, swallows reservoirs and contributes to the reduction of water supply and quality from the Cidanau catchment.

PES as One Initiative to Rehabilitate the Cidanau Watershed

The numerous efforts that have been made to overcome the watershed problems in the Cidanau have had limited success. These include a transmigration programme for the communities living in the Rawa Danau area, reforestation, and land rehabilitation activities. Key issues in the failure of past efforts included a lack of consultation and joint planning between key stakeholders and a lack of attention to social outcomes.

The failures of these previous efforts at watershed management triggered a group of people concerned about the degradation of Rawa Danau to establish the FKDC in 1998. The forum tried to increase awareness among the public and within local government about environmental problems and integrated watershed management by conducting seminars and discussions. This forum received recognition from the newly established Banten provincial government[3] and gained legal status through a Governor's Decree in 2002.

The concept of payment for watershed services in Cidanau was

Table 5.3   The stakeholders involved in the PES scheme

Role	Stakeholders
ES Providers	Four upstream farmer Groups from Cidanau (Citaman, Cibojong, Kadu Agung villages)
ES Buyers	Current single buyer: PT KTI Potential buyers: other companies in Cilegon such as PDAM (state-owned water company), Krakatau Steel, Ronn & Hass, PT Pelindo, PT Politrima, Chandra Asri, Bakrie Group
ES Intermediaries	Forum Komunikasi Cidanau (FKDC) – a multi stakeholder forum
Policy makers	District government and legislative officers of Serang (upstream) and Cilegon (downstream) Provincial government and legislative officers of Banten National watershed management body coordinated by the Ministry of Forestry
Main supporting NGOs	Rekonvasi Bhumi, LP3ES
Main supporting university	Bogor Agricultural University
Main supporting international agencies	World Agroforestry Centre, International Institute for Environment and Development, GTZ

introduced by international organizations, such as *Deutsche Gesellschaft für Technische Zusammenarbeit* (German Technical Cooperation, GTZ), the World Agroforestry Centre and the International Institute for Environment and Development in 2002. A member of *Rekonvasi Bhumi* (a local NGO) visited Costa Rica to see the implementation of a PES programme funded by GTZ. The conditionality aspect, the involvement of multiple stakeholders in watershed management and the innovative nature of the Costa Rican PES scheme stimulated interest to trial such a scheme in Cidanau. In 2004, the FKDC invited the PT Krakatau Tirta Industry (PT KTI) to join this scheme and started facilitating negotiations between private land-owners in the upper watershed and the company.

The Stakeholders, their Roles and Responsibilities

The PES scheme involves many stakeholders, including farmer groups, downstream companies, government officers from district, provincial and national levels, supporting NGOs and universities (Table 5.3).

Table 5.4   Farmers involved in the PES scheme

Village	Number of farmers	Starting year
Cikumbuen	32	2007
Citaman	43	2005
Kadu Agung	38	2007
Cibojong	29	2005 (ended after 2 years)
Total	142	

The sellers of the environmental service
In total, 142 farmers were involved in the PES scheme: 43 from Citaman, 29 from Cibojong, 38 farmers in Kadu Agung and 32 in Cikumbuen (Table 5.4). Participating villages were selected according to the mapping of critical land by the local government (for example, steep slopes and erosion-prone soil) and participating farmers at each village were selected by considering their involvement in farmer groups and private ownership. Aside from land ownership, no other socioeconomic criteria were considered, as the intermediary felt that there was relatively equal wealth distribution and landownership rates among the communities, with the typical land of each household being between 0.2 and 0.5 ha.

The buyer of the environmental service
KTI – the only authorized company managing water from the Cidanau watershed – is the only buyer in the current PES scheme. The water from upstream flows through a 28 km pipe to the water treatment reservoir. KTI initially used this clean water for its steel industry operations. Recently, this company has also been supplying about 80 per cent of the water needs of 120 companies at Cilegon, such as PDAM (a state-owned company that supplies drinking water, which purchases the water at a subsidized price), and Indonesia Power Company, which supplies electricity to Java and Bali. This highlights the importance of the Cidanau watershed for industrial activities. KTI clarified that the initial source of funds for the PES scheme came from the operational budget of the company, and PES funding was drawn from corporate social responsibility funds.[4] The company's staff remarked that the motivation for engagement in PES was to support conservation efforts in the Cidanau watershed, rather than securing access to clean water for the production process. The company's staff mentioned that the government was the one responsible for the maintenance of the constant flow of water.

The intermediary for the environmental service

FKDC's role in the PES scheme is to manage funds, to facilitate contracts with farmer groups, and to monitor and verify rehabilitation activities. Its additional role is to raise awareness of payment for environmental services amongst other potential buyers in the Cilegon industrial area. FKDC added an *ad hoc* team within its structure in 2005 specifically to facilitate the scheme. This *ad hoc* team consists of representatives of government institutions at the provincial and regency levels in the Cidanau watershed area and an NGO.

This team plays an intermediary role by: (1) managing the payment of PES funds from the buyer to the farmers for their rehabilitation and conservation activities; (2) supporting planting activities on private farms involved in the PES project; (3) encouraging other potential buyers to join the scheme; and (4) advocating the integration of the PES scheme in the provincial and district governments' environmental management policy.

Setting the Price for the Environmental Service

The price-setting process in Cidanau was based on negotiations between the buyer (KTI), the intermediary (FKDC) and the sellers (farmer groups). The agreed price was formalized in a Memorandum of Agreement between KTI and FKDC (represented by the Governor of Banten Province). After this agreement, the chairperson of the FKDC *Ad Hoc* team and farmers' groups from Citaman and Cibojong made another agreement covering a total land area in two villages of 50 ha. In 2007, the other two villages (Kadu Agung and Cikumbuen) joined the initiative, each with 25 ha.

The annual rate set in the contract between the KTI and the FKDC was US$350[5] per hectare based on input costs, calculated according to funding levels provided in government tree-planting programmes (land preparation, ground cover, seedlings, transport, fertilizers and labour) on state lands. The market value was established by referring to the cost per hectare of the national forest rehabilitation programme (GERHAN) coordinated by the national government. KTI agreed to make three payments within 5 years, and were subject to a 6 per cent tax. The total payment of the KTI to the FKDC was US$35 000 for Phase 1: 2005–07 and US$40 000 for the following Phase 2: 2007–09. The payment for the fifth year was to be renegotiated.

The *Ad Hoc* Team initially offered farmers annual payments of US$75 per ha. The annual payments were agreed at US$120 per ha, provided that 500 trees per ha were planted and plantings maintained. The FKDC scaled down the payment to farmers in order to cover all the five-year payments with the available four-year funds from KTI – or, in other words, to provide a buffer

Table 5.5   Actual allocation of revenues by the FKDC in the first four years*

Allocation	US$	Proportion of total payments made (%)
Payments to farmers	60 000	80

Transaction cost
• 40% for conducting capacity building and searching more buyers (dissemination, publication, seminars, etc.)
• 27% for monitoring and verifying field activities
• 33% for operational cost:
  – 16% for paying personnel costs
  – 11% for organizing meetings
  – 6% for administration
	10 500	14

Tax	4 500	6
Total	75 000*	100

Note: * This amount is the payment from KTI for Phase 1 and Phase 2 (4 years). KTI still has to transfer the remaining funds for the fifth year, as much as $US100 000, which is contingent on current performance. The total commitment should be $US175 000 (100 ha × $US350 per ha × 5 years).

in case KTI did not meet its obligations. They took this risk-management action because they still have to negotiate the fifth-year payment in 2011. From the interview with the FKDC members, they plan either to involve new farmer groups in other villages or to extend the contract with the current farmers if the KTI disburses its third payment in 2011.

Payment Allocation

Since it had a key role in the agreement and disbursement of payments to farmer groups, FKDC took responsibility for managing many of the transaction costs for buyers (Table 5.5). FKDC members estimated that the transaction cost was around 14 per cent of the annual payment, including the costs of capacity-building activities, searching and contacting new buyers, information dissemination, and monitoring and verifying performance of agreements in the field.

Farmers used about 95 per cent of their initial payment to buy seedlings, plant and maintain the trees, and were left with around 5 per cent to spend on their own priorities, including investment in local business in their first

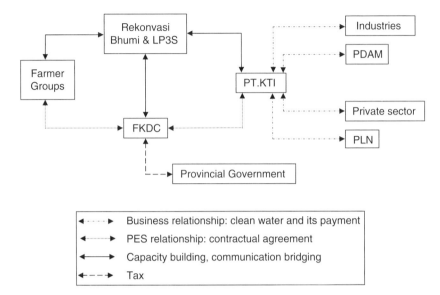

Notes: FKDC – Forum Komunikasi DAS Cidanau (Communication Forum of Cidanau Watershed); PDAM – state-owned drinking water company; PLN – state-owned electricity company.

Source: Adapted from Budhi et al. 2008.

Figure 5.1   The PES scheme relationship and flows of services

year. Interviews indicate that the operational costs for the second year were 50 per cent lower, and many farmers chose to invest the balance in their businesses.

The contract between the FKDC and the farmer groups in the four villages involves a yearly payment of US$120 per ha for 5 years, subject to satisfactory implementation of the rehabilitation works activities, including planting and maintaining timber and fruit trees at a minimum of 500 trees per ha and no cutting during the contract period.

During the first year of the contract, 30 per cent is paid on the signing of the contract, 30 per cent is paid after six months of implementation and the remaining 40 per cent is paid at the end of the year. Subsequent annual payments of US$120 per hectare are made for the next 4 years (40 per cent is disbursed in June and 60 per cent in December), subject to the satisfactory implementation of the rehabilitation works.

All members of the first two farmer groups received their first payment in May 2005. Three months later, the FKDC commenced monitoring and requested records of tree-planting on contracted lands. In Citaman, the *Ad Hoc* team found that 0.5 ha was not being maintained as per the agreement because the owner left the village for a new job. However, since the other members of the farmer group had accomplished the minimum requirements of the contract, the *Ad Hoc* team did not disqualify the group. The group decided to manage the 0.5 ha of land and charged the owner the operational costs of managing this land under the contract. The contract is a collective one. If a farmer breaks the rules, the *Ad Hoc* team will terminate the contract of all the members. The collective contract was chosen over individual contracts because the team assumed that by applying the 'sharing responsibility principle',[6] it could strengthen internal relationships and self-monitoring among group members.

Implementation Problems

A number of issues associated with the PES scheme were raised by FKDC members in focus group discussions.

First, the FKDC found it difficult to communicate the unique characteristics of an incentive-based mechanism to other stakeholders, such as local governments and buyers, because of their relative inexperience with the operation of such mechanisms. Buyers often viewed the scheme as adding another layer to their operational costs and have, in many cases, used corporate social responsibility funds to cover the ES payment (which means it is accounted for as a promotional rather than an operational cost). Second, lengthy negotiations were unavoidable given the number of stakeholders involved and because KTI was unwilling to pay the farmer groups directly. There were at least three stages of negotiation over 2 years. The first stage, to establish the main design elements, was between the *Rekonvasi Bhumi* and the *Ad Hoc* team of FKDC and took a period of 8 months. The second negotiation period, to draft the contract, between the KTI and the *Ad Hoc* team lasted about 6 months. The third negotiation phase, to develop contracts for payment amounts and conditionality, was between the *Ad Hoc* Team and farmer groups in the villages of Citaman and Cibojong.

Third, FKDC members expected the communities to have a more active role in conserving the watershed rather than depending on the PES payment for any environmental conservation. There was confusion about whether any formal regulation by provincial government could play an important role in targeting more ES buyers as well as providing an enabling policy environment with strong political support. They stated that such regulations were needed, but did not have ideas about the contents of such regulations. Without the certainty of voluntary participation of additional buyers, FKDC was less able to encourage more sellers to engage in the scheme. Meanwhile, KTI demanded regulations obliging potential buyers to participate in the PES, having assumed that such regulations would optimize the role of additional buyers in conserving the watershed.

After 2 years of implementation, the Cibojong village did not achieve the target stipulated in the contract and the contract was terminated. A farmer cut the trees on about 0.14 ha of land, reporting that the trees had been stolen (an investigation later found out that one of his family members had cut the trees to buy a motorcycle). Procedurally, a report should have been made to the FKDC, together with a letter from the police department guaranteeing that they would not breach the contract further. However, this was not done, and the members assumed that the contract had been cancelled. Villagers continued to cut trees on the PES-contracted lands, based on their assumption that the scheme would not provide them any further payment. An interview conducted by the FKDC with the members revealed that most would have preferred to remain in the scheme. Therefore, the cancelling of the contract would likely have been avoided if the group had advised the members of the correct procedure following the initial (illegal) cutting of the trees.

THE IMPACTS OF THE PES SCHEME

The Environment

A clear assessment of the environmental outcome of the scheme is not available yet. Although some data were presented earlier in this chapter on decreases in water quantity and quality in Cidanau, the actual link between the land-use practices used to promote watershed protection and water supply are unclear. Also the scale of the current PES scheme may have been limited in its environmental impact given the size of the watershed. The monitoring system for the scheme relied on the accomplishment of contractually agreed land-use practices as a proxy for environmental outcomes. FKDC members, particularly those from KTI, have visually observed that the water supply is relatively stable in 2008, but so far this has not been backed up by scientific evidence.

The Livelihoods of the Participants and Non-participants

Financial capital
According to FGDs, the communities in Cidanau earn their income from the tree-crops – *melinjo*,[7] coconut, robusta coffee, durian and clove – which represent the top five income sources, and further planting of these tree crops was supported through the PES scheme. The FGDs did not indicate

Table 5.6   Household income sources

Source of income (%)	After PES (2005–now)		Before PES (2000–05)		Before PES (before 2000)	
	P	NP	P	NP	P	NP
Melinjo	27	28	23	32	15	17
Farming labour	15	15	0	8	0	13
Coconut	12	8	10	8	15	10
Clove	10	7	18	7	12	10
Coffee	10	10	15	10	17	18
Durian	7	3	13	8	23	12
Salak[a]	5	8	5	5	3	0
Wood	5	7	8	0	0	0
Payment for ES	3	0	0	0	0	0
Banana	2	2	3	3	3	12
Cocoa	2	0	0	0	0	0
Petai[b]	2	7	0	5	0	0
Cotton	2	0	3	2	5	2
Jengkol[c]	0	0	0	0	5	0
Paddy	0	0	0	0	0	2
Upland paddy	0	2	0	5	2	0
Others (clove labour, livestock labour, motorbike renting, construction labour, trader)	0	3	0	7	0	5

Notes: Totals may not add to 100 due to rounding.
P for participants and NP for non-participants.
[a] *Salak* (*Salacca zalacca*) is a cultivated palm tree native to Indonesia and Malaysia. The fruit is also known as snake fruit because of its reddish-brown scaly skin. The pulp of the fruit is edible with a sweet and acidic taste.
[b] *Petai* (*Parkia speciosa*) is a leguminous tree; its beans are an acquired taste and are combined with other strongly flavoured foods for traditional Asian dishes.
[c] *Jengkol* (*Archidendron pauciflorum*) is also a leguminous tree. Similar to *petai*, the beans are frequently served as a vegetable salad or accompaniment to Indonesian dishes.

Source: Adapted from Budhi et al. 2008.

significant changes in income sources between the periods before 2000, 2000–2005 and after the introduction of the PES in 2005 for both participants and non-participants (Table 5.6). Tree species were selected on the basis of commodity prices and market demand to enable participants to build their productive base of valuable tree crops.

Most of the participants reported *melinjo* to be one of their main income sources before and after the PES. The farmer groups in Citaman have also invested some of their PES income to develop a *melinjo* nursery, with training on processing *melinjo* into crackers from the Rekonvasi Bhumi. These value-adding activities might enable communities to gain a better return from *melinjo*. Some participants stated that wage labour from farming and other sectors, such as construction and business (for example, motorbike rental), also contributed more to their household incomes in recent years compared with agricultural products. The communities have become more dependent on labour income compared with income from agriculture because most of them have sold their lands, or only had small land areas to begin with, which could not fulfil their income needs from agriculture. In addition, the PES contract constrained the clearing of lands which participants owned, and respondents added that this gave them more time to undertake alternative work, such as paid labour.

The indications are that the PES contract in Cidanau did not have a major impact on the livelihood options pursued by communities because of their existing reliance on tree crops as a primary income source before the scheme commenced. Some participants did mention, however, that they had lost income from wood harvesting and wanted the option of continuing with tree thinning on their contracted gardens. The income from the wood harvest could be as high as US$200 annually, around 60 per cent higher than the value of the PES contract. Wood harvesting had previously contributed an estimated 5 to 7 per cent of household income for both participants and non-participants. The discussion also recorded the use of some 40 types of commodities, including leaves, flowers and fruits that are locally marketable.

The annual PES income of US$120 per hectare contributed only around 3 per cent to PES participants' household incomes. Only one group in Citaman regarded PES as a primary source of income. The rest considered PES income to be short term and not a primary livelihood source, although during the 4-year operation of the scheme the total payment might have exceeded their income from selling fruits. Around half of the participants assumed that the PES contract could increase the price of their land, although most non-participants did not consider it likely that the land price would rise as a result of the PES scheme. No transaction on land allocated to the PES scheme has occurred, therefore there is no information about the impact on the value of land.

The PES scheme has stimulated local business, mostly because of additional business development support from NGOs and government agencies involved in the PES scheme. The facilitating NGO *Rekonvasi Bhumi* (together with the Serang Service Office of Industry, Trade and

Cooperatives) has supported farmer groups with entrepreneurship and marketing training, and also gained advice on technical issues from the Environment Technology Agency (Munawir and Vermeulen 2007). Some areas of local business development have included the production and marketing of vegetable oil from *nilam* (*Pogostemon cablin*) and *melinjo* cracker production. FKDC members had observed that the PES scheme provided a locus for greater government support to the participating villages to: (1) establish a nursery of fruit trees; (2) develop local business for edible mushrooms in Citaman and Kadu Agung; and (3) establish a poultry project in Cikumbuen. They felt that the reputation of these villages had been raised due to their participation in the PES scheme.

Human capital
PES participants and non-participants attended occasional training conducted by the Agricultural Service and Forestry Service of the local government, dealing with coffee, *melinjo*, timber and fruit tree cultivation. However, the PES scheme had a particular impact on the capacity, skills and knowledge of participants (Table 5.7) because of their regular interaction with NGO staff and researchers. PES participants were more aware of environmental issues, such as the causes of erosion, land slides and downstream sedimentation, as well as management measures, such as erosion prevention, prevention of illegal cutting of trees, waste management, and the role of trees in water and soil conservation. However, only about 30 per cent of the participants and 17 per cent of the non-participants knew about the concept of PES and how the value of the contract could be calculated.

PES participants also reported improved capacity and skills in managing the farmers' organization, including networking to improve local business and to improve implementation of the PES scheme. This capacity-building occurred through interaction with the FKDC members.

As noted earlier, some participants observed that they had more available time and less activity on their lands due to restrictions on activities under the PES scheme. Because of this, PES participants and non-participants focus groups identified a need for training in alternative livelihoods, such as (1) raising livestock and poultry; (2) cultivating fruit and timber trees; (3) making fruit crackers, from *melinjo*, banana and cassava; (4) pest management; (5) establishing freshwater fish ponds; (6) apiary businesses; and (6) cultivating mushrooms. Women identified an interest in training in literacy, sewing and cooking. The FKDC members added that the communities also might need further training to strengthen their local institutions.

Interviews with the FKDC members indicated that their knowledge

Table 5.7   Type of knowledge/capacity/skills gained by participants and non-participants after the PES implementation

Type of knowledge/capacity/skills	Participant (%)	Non-participant (%)
Conservation		
Causes of erosion, land slides and downstream sedimentation	100	17
How to maintain clean water and to reduce air pollution	83	–
Roles of trees in conservation	67	–
Simple construction to prevent erosion	50	–
Understanding of PES concept	33	17
Institution and governance		
Ability to govern an organization	67	17
Ability to solve problems within farmer groups	67	–
Administration of farmer groups	50	17
Networking to improve local business and PES implementation	50	–
Transparent financial management	33	–
How to develop local business		
Livestock	33	17
Agriculture	17	–
Fishery	–	–

about PES issues increased, such as the principles of PES, how to design community-based forest management, how to strengthen local institutions, global issues such as global warming, the Clean Development Mechanism, and Reducing Emissions from Deforestation and forest Degradation.

Social capital

Aspects of social capital discussed in communities include behavioural norms within the community, reciprocity between community members, trust and the existence of internal and external networks, before and after the implementation of the PES scheme.

The focus groups with PES participants in Citaman revealed that they had written rules to guide members of their farmers' group towards meeting their collective obligations under the PES contract: if one member defaulted on the agreement, this would become the responsibility of the whole group. Sanctions would be imposed on such a member in the form of expulsion from the group. In other villages, there were no written rules, but people knew the rule that trees should not be cut in the contracted areas.

The sanction for cutting trees involved a police report, as well as informal social sanctions at the community level. The informal sanctions included exclusion from social gatherings. The participants also commented on rent-seeking by local government staff in relation to PES payments, that is, requesting part of the payment for contributing to village income.

All the participants that joined the focus groups knew about the written contract between their group and the FKDC, and that observing restrictions on cutting trees was necessary to receive payments, while cutting trees would lead to contract termination. Some participants observed that the local NGO, *Rekonvasi Bhumi*, used informal warnings as the first step if contract infringements occurred. (Farmers from Cibojong village, where the contract had been cancelled, were not participants in these focus groups.)

The PES contract brought opportunities for participating communities to interact more with other external stakeholders, which expanded the external networks of these communities to include: (1) researchers conducting studies on PES in Cidanau; (2) local NGOs who facilitated the PES contract; (3) the KTI as the buyers; (4) the FKDC as the intermediary; (5) other government agencies besides the Agriculture and Forestry Services, such as the Natural Resource Service. In contrast, non-participants only mentioned increased interaction with the local NGO and government agencies amongst their new contacts after PES.

The focus groups discussed issues of trust within the community and between community members and external stakeholders (Table 5.8). Trust was seen as the ability to receive and give assistance from people beyond the immediate household and relatives in case of shortness of money or food. Focus groups reported that trust amongst community members (both participants and non-participants) in Cidanau was relatively high, while the level of trust between community members and external stakeholders was lower. This is consistent with the observation that the four villages involved in the programme have a high degree of internal homogeneity. Most of them are Moslem and their wealth strata are almost equal, which may contribute to ease of interaction and trust.[8] In Cidanau, communities usually participate in regular collective action events to produce public goods and services, such as maintaining roads, bridges, community buildings and water supply systems. These activities are an important aspect of rural social capital in Indonesia (Grooteart 1999). This also appears to be the case in Cidanau.

Some key persons, mostly group chairpersons and village elders, lead in negotiations with external stakeholders and gain access to more information than other participants. There were some signs of jealousy amongst non-participants about their exclusion from the PES scheme (which was

Table 5.8   Trust among internal and external stakeholders

Relationship	How trust is expressed
Amongst participants	Borrowing money and rice Sharing information Mortgaging (loans) Collective labour sharing
Participants and government	Making identification and family card Paying tax Receiving administrative information Getting cash assistance* Maintaining security
Participants and non-participants	Collective labour sharing Sharing information Borrowing money, rice, daily needs and construction materials
Participants and FKDC	Delivering the payments for accomplishing the contracts Sharing information Maintaining transparency in managing the funds of organizations
Participants and the state plantation company PERHUTANI	Giving seedlings Giving information Giving access to manage forest and plant ally-cropping on the area of PERHUTANI
Participants and NGO	Implementing programmes Sharing information, especially on environmental services Conducting meetings

Notes: * The Indonesian government has a programme called *Bantuan Langsung Tunai,*or direct cash assistance, as one of its programmes for buffering the poor from the financial crisis.

the result of the limited budget of the buyer). The interaction between participants and non-participants in the same village decreased as the interaction between participants and other external stakeholders increased. This condition somehow created an exclusive group of PES participants who did not mix socially with other villagers. The FKDC members also mentioned this tendency.

There was general agreement that trust between communities and government was lower after 2000 and has become worse since the start of the PES project. The communities do not consider the government a partner from whom they can ask for assistance. The communities felt a reduced level of confidence in the government's capacity and commitment to provide public services (Table 5.8). Since 1998, Indonesia has been in a period of transition known as '*Reformasi*' (Reform in Indonesian). Although this period has been characterized by greater freedom of speech, many rural communities considered that they had more secure livelihoods during the earlier Suharto-dominated period, which involved unprecedented national growth and greater integration of rural areas into national development. The *Reformasi* era provided greater autonomy to village level governments. However, there have been fewer nationwide programmes, as local conditions vary greatly and severe financial constraints during 1997–1998 led to reduced government spending on rural development (Antlov 2003). The communities in Cidanau noted that the government had paid less attention to rural development after the beginning of the *Reformasi* era and felt a diminished sense of trust in the government. *Rekonvasi Bhumi*, the only NGO that is active in advocating the PES concept, was established soon after the beginning of the *Reformasi* era, when greater space was created for civil society. In Cidanau, interaction between community members and this local NGO nurtured a level of trust; the same was true with FKDC, the ES buyer.

Government officials shared the view that the existence of the PES scheme had increased their communication with stakeholders such as the FKDC members and the KTI, as well as a need for greater inter-agency communication. They expected that PES could assist the government in conducting its conservation programmes and in improving the communities' livelihood.

Natural capital
Since the PES scheme only targeted individual farmers, and restrictions on land use only applied to private lands, there was no change in access to common resources. Before the scheme and after its beginning, communities in Cidanau utilized non-timber products from the forest, such as water, wild boar, fish, firewood, medicinal plants, herbs, fruits and leaves. Around half of the participants did comment, however, that the PES contract had reduced their access to timber for construction because they could not harvest the timber from the contracted land. Currently, they have to buy some wood to fulfil their own needs. The FKDC reported that, at the end of the contract, the farmers would be allowed to cut 40 per cent of their current plantings to fulfil their needs for wood and increase their incomes if they are willing to continue the PES contract.

Both participants and non-participants knew the benefits of maintaining natural resources. They could explain environmental services provided by the healthy ecosystem and claimed that they had had this knowledge for a long time. According to informants, the services provided by an intact watershed and the Rawa Danau Reserve included providing timber for construction and non-timber forest products, storing water, avoiding floods, landslide and erosion, contribution to a comfortable micro-climate, fertilizing soils and ecotourism, particularly for the Rawa Danau. In addition, the local government and the buyer added that the Cidanau watershed had high and strategic economic value because it supported the existence of important industries and households in the towns of Cilegon and Serang.

The communities have been involved in various rehabilitation activi-ties (both government-initiated and locally organized) before and after the PES scheme. Government programmes included planting trees, such as mahogany, clove, *albizia* and *calliandra*, joining forest fire prevention activities and forest patrols for the prevention of illegal logging, and ter-racing steep lands. The Cidanau communities were also involved in the National Movement of Land Rehabilitation. Self-supporting activities included cleaning the river annually in Kadu Agung and planting bamboo and productive trees, such as *melinjo*, durian and stink bean. However, these actions are mostly patchy, not integrated and short-term with uncertain success.[9] In addition, the PES project did not set up systematic monitoring for environmental services in Cidanau. The KTI claimed that the sedimentation and water quality in Cidanau had improved in the last 2 years. However, whether this conclusion is correct, and whether the change in ES has any connection with the PES scheme, have not been scientifically demonstrated.

Physical capital
The Citaman group invested 5 per cent of its PES payments to build a 100 m pipeline for clean water to serve about 50 households. This water pipeline also served non-participants, but they were required to pay a service fee of US$0.30 per month or 1 kilogram of rice. The Kadu Agung group planned to build a village mosque from all funds collected through the PES contract. Other villages did not report plans to invest their money in education and health improvements. Their investments in physical capital were a collective decision driven by their specific needs. Villages without any investment plans might simply not have collective needs.

Participants in focus groups complained about the poor condition of roads, which doubled their transportation costs. This has been the case for many years and a change of government did not bring any changes to their village assets. However, the discussions with the FKDC highlighted the fact that the community had received assistance to develop a nursery and a building for community meetings in Ciomas village. The budget for these activities came from the provincial government in 2005 because it noticed the existence of PES activities in the village.

The FKDC has no further plans to develop public facilities in the villages covered by the PES scheme. Nevertheless, it agreed that developing public infrastructure in the sellers' villages could multiply the positive impacts of the PES scheme. For example, better roads to the villages would increase accessibility and improve communication, coordination and monitoring as well as contributing to wider economic and social development.

CONCLUSION

Livelihood Impacts

The Cidanau PES scheme has affected the livelihood of PES participants and non-participants. Benefits were mostly non-financial: expanded social networks with external stakeholders; knowledge and capacity of the community; and small-scale public infrastructure investments. Direct financial benefits were limited. So far, four villages out of five have proved successful in meeting the contract terms; however, there is a need to investigate further whether the non-financial benefits and limited financial benefits are sufficient to cover their 'total opportunity cost'. We presume that these benefits, combined with recognition from governments and external stakeholders, can increase farmers' commitment to the scheme. It is important to adjust the value of the new contract so that farmers can cover their true opportunity cost if the funds from the buyer allows that. This finding is in line with the conclusions in other PES sites in Asia (Leimona et al. 2009).

Although the PES scheme did not drastically change the livelihoods of participants, linkages with external stakeholders had begun to create options for participants to diversify or capture greater value from their income sources. The external stakeholders are largely partners in the PES scheme, such as the FKDC and a local NGO. Exposure to these partners also increased the participants' knowledge of conservation, as well as the skills to manage the farmers' organization, and helped to build networks to improve their businesses and implementation of the PES scheme.

Participants and non-participants reported that they were aware of the benefits of conservation before the PES scheme was implemented. Their understanding of the PES concept was still limited. The capacity-building for the PES concept at the local level has been important. However, future capacity-building should also be focused on tangible aspects of the PES scheme and problems that put barriers at the local level in implementing PES, such as lack of information about good planting materials and knowledge of tree management.

The PES scheme has created new standards and mechanisms for managing behaviour around natural resources. It supports the establishment of new written and unwritten rules as well as sanctions related to natural resource management and land-use practices. The PES contract sets out formal rules and sanctions binding the sellers and the intermediary which supplement their existing informal rules and sanctions. These informal rules and sanctions were useful to support collective action and induce the accomplishment rate of the PES contract.

There were signs of jealousy among non-participants in Cidanau towards the participants due to their exclusion from the PES scheme. So far, such jealousy has not destroyed social relationships in communities because the size of payments is limited and it has not created inequality. The investment of PES income in community infrastructure, such as water supply, mosques and meeting halls might reduce social conflict as such investments extend to the indirect beneficiaries of the scheme, although not to the same degree in some cases. Improved government investment in PES villages, as planned but yet to be implemented, could also help to reduce the risk of potential conflict between participants and non-participants.

Access to common pool resources, such as state forests, did not change with the implementation of the PES scheme because only non-timber products were taken from the forest.[10] However, the restrictions posed by the PES scheme on landowners' access to timber on their own lands could lead to illegal logging on common lands – that is, it could result in so called 'leakage'. Monitoring of the nearby environment should therefore be carried out by the PES scheme.

Environmental Impact

There is insufficient scientific evidence to judge the impacts of the Cidanau scheme on environmental services. Although the selection of contracted villages was based on criteria (steep slopes and erosion-prone soil) that would maximize environmental outcomes, and stakeholders in the scheme believing that planting trees would solve the watershed problems in Cidanau, the causal link between changing land-use practices and increasing ES are unclear and indirect. For the next step, identifying and monitoring specific indicators of watershed services in Cidanau is crucial. For instance, a rapid hydrological assessment in Singkarak, West Sumatra, Indonesia (Farida et al. 2005; Jeanes et al. 2006), concluded that raising the water level of the lake, sought by the ES buyer to increase its hydro-electric performance, is mostly influenced by changes in mean annual rainfall and only mildly by land cover. Without an understanding of watershed functions, and related indicators, PES schemes such as this may not achieve the desired environmental impact, leading to disappointment amongst sellers and buyers.

Design of the PES Scheme

The amount of the payment per hectare set out in the PES scheme in Cidanau was based on input costs for tree planting. Information on opportunity costs is not yet available for Cidanau. Farmers might have accepted the contract without further consideration of the real costs and benefits involved in the scheme. The agreed value of the contract might not fully represent the real opportunity cost of the farmers because of the dominant position of the intermediary. The transaction cost in Cidanau was about 14 per cent of the total payment.

In terms of lessons for REDD, the Cidanau case raises important issues about the need to factor in opportunity costs when negotiating payments to ensure their long-term sustainability. It also highlights the need for aware-ness of the social dynamic between participants and non-participants and the need to design benefit packages to minimize community-level conflict. The Cidanau case suggests that the role of the intermediary is very impor-tant and possibly dominant. An honest and trusted intermediary is one of the keys to success.

NOTES

1. The 64 members of this forum are upstream and downstream stakeholders. The upstream stakeholders include farmer groups, the government of Serang district, the Serang legislative body, provincial agriculture services (provincial and district forestry and environment), provincial and district planning agencies (BAPPEDA), provincial human capacity and development agencies, provincial human settlement and regional infrastructure services and a non-government organization. Downstream stakehold-ers include representatives of the PT Krakatau Tirta Industry (KTI) (a private water company), the government and legislative body of Cilegon District, agriculture serv-ices and urban water users. This body was later to become the primary coordination mechanism for PES.
2. The land cover of the Cidanau watershed is dominated by agriculture (71 per cent): consisting of mixed farming (36.7 per cent) and rice fields (34.4 per cent). The remaining land cover is forest (18.5 per cent) and swamp forest (8.4 per cent) (Adi 2003).
3. Banten was a district in West Java Province before 2000 and became a new province in 2000.
4. In Indonesia, a state-owned company must allocate 1 per cent of the net benefit of state-owned companies to develop environmental programmes with communities. The legal

basis of this scheme is the Letter of Ministry of State-owned Company Affairs about Corporate Social Responsibility Partnership Program (KEP-236/MBU/2003).
5. $US1 = Rp10000.
6. In Indonesian, the term is *tanggung renteng* literally meaning an individual failure will become collective failure.
7. *Melinjo* (*Gnetum gnemon*) is a plant native to Indonesia. The seeds are used for vegetable soup, or ground into flour and deep-fried as crackers.
8. Rahadian, Director of *Rekonvasi Bhumi*, pers. comm. (2008).
9. Reports on the failure of the National Movement of Land Rehabilitation are numerous (see, for example, As 2006). One of the reasons for this failure is that the programme is top-down with very little participation from the community. The government dominates the supply of the plant materials and determines the species that should be planted. The community acts as labourers for the planting activities and they are often not interested in maintaining their plantations because, in some cases, they do not have access to the harvest.
10. Further investigation on this should be done because some literature mentioned that deforestation had been a big problem in Cidanau (Kiely 2005).

REFERENCES

Adi, S. (2003), *Proposed Soil and Water Conservation Strategies for Lake Rawa Danau, West Java, Indonesia*, Water Resources System, Hydrological Risk, Management and Development, International Association of Hydrological Sciences Publication No. 281, Wallingford, UK: IAHS.

Antlov, H. (2003), Village government and rural development in Indonesia: the new democratic framework, *Bulletin of Indonesian Economic Studies* **39** (2), 193–214.

As, M. (2006), GERHAN Gagal? *WARTA Forum Komunikasi Kehutanan Masyarakat*, available at http://www.fkkm.org/Warta/index2.php?terbitan=no e&action=detail5&page=17 (accessed 13 November 2009).

Budhi, G.S., S.A. Kuswanto and I. Muhammad (2008), Concept and implementation of PES program in the Cidanau watershed: a lesson learned for future Environmental Policy, *Policy Analysis of Farming* **6** (1), 37–55.

Darmawan, A., S. Tsuyuki and L.B. Prasetyo (2005), *Analysis on extension of illegal cultivations from socioeconomic aspects in Rawa Danau Natural Reserve, Banten, Indonesia*, available at www.geocities.com/rubrd_grup_1/. . ./O1_2_arief_darmawan.pdf (accessed on 19 September 2009).

Farida, J.K., D. Kurniasari, A. Widayati, A. Ekadinata, D.P. Hadi, L. Joshi, D. Suyamto and M. van Noordwijk (2005), *Rapid Hydrological Appraisal (RHA) of Singkarak Lake in the Context of Rewarding Upload Poor for Environment Services (RUPES)*, Bogor: World Agroforestry Centre.

FKDC (2005), *Forum Komunikasi Das Cidanau. Propinsi Banten (with English Version)*, Publisher Unknown, Indonesia.

Grootaert, C. (1999), *Local institutions and service delivery*, Local Level Institution Working Paper No. 5, Washington DC: The World Bank.

Jeanes, K., M. van Noordwijk, L. Joshi, A. Widayati, J.K. Farida and B. Leimona (2006), *Rapid Hydrological Appraisal in the Context of Environment Service Rewards*, Bogor: World Agroforestry Centre.

Kiely, A. (2005), Can agricultural intensification address the problem of deforestation in Cidanau Watershed, Banten, Indonesia? A basis for the allocation of environmental service payments, MSc thesis, Imperial College London, UK.

Leimona, B., L. Joshi and M. van Noordwijk (2009), Can rewards for environmental services benefit the poor? Lessons from Asia, *International Journal of the Commons* **3** (1), 82–107.

Munawir, S. and S. Vermeulen (2007), *Fair Deals for Watershed Services in Indonesia,* London: International Institute for Environment and Development.

Rekonvasi Bhumi (1999), *Laporan Hasil Perjalanan Observasi Rawa Danau 30 Januari 1999,* unpublished report prepared for LSM Rekonvasi Bhumi.

6. The 'No-Fire Bonus' scheme in Mountain Province, Cordillera Administrative Region, Philippines

Rowena Soriaga and Dallay Annawi

INTRODUCTION

The incidence and scale of forest fires are increasing in Southeast Asia. Apart from the contribution to climate change that occurs when forests turn from carbon sinks into carbon sources, forest fire also has significant social and economic impacts (FAO 2007). Although fire has a legitimate place in traditional agriculture and has been an important factor in the development of terrestrial ecosystems, concerns about its social and environmental impacts are growing (Walpole et al. 1993, 1994; Karki 2002; FAO 2006). As climate change is predicted to increase the frequency, persistence and magnitude of drought in the region's tropical forests, the significance of fire management in attempts to reduce emissions from forestry can only increase. This case study from the Philippines reviews key design factors and impacts in a payment scheme to reduce fire incidence through local engagement.

In 1996, the 'No-Fire Bonus' (NFB) scheme was launched in the fire-prone pine forests of Mountain Province, part of the Cordillera Administrative Region in northern Philippines and the ancestral domain of several indigenous groups. While the scheme involved payments for fire protection, it was not developed with attention to PES principles, where a clearly defined environmental service is conditionally traded between providers and buyers, as PES is largely a nascent concept in the Philippines. Nevertheless, the case provides important lessons about the operation and impacts of incentive payments for fire protection that are transferable to REDD schemes. In particular, it highlights the role of secure tenure in securing buy-in and benefits for indigenous peoples living in the area, and the important role of local government in facilitating the scheme.

This chapter starts with a background section on the NFB scheme and its implementation. The livelihood and environmental impacts of the scheme are analysed as a basis for identifying relevant lessons for the viability and sustainability of such schemes. The study is based primarily on two visits to the case study site.[1] Key informant interviews were conducted in seven *barangays* (villages) in two municipalities with (former and current) *barangay* officials, community elders, farmers, women, municipal government officials and forest department field staff.[2] Group discussions on the scheme were undertaken with *barangay* government officials during a larger municipal government workshop to launch a national reforestation programme. Fire investigation reports from the local, provincial and regional levels that were available were reviewed to gain an understanding of fire trends during and after the scheme's operation.

Data limitations prevent this study from directly attributing the social and environmental changes observed to the *barangay*-level scheme. First, there is a lack of *barangay*-level baseline information to show fire trends before implementation. Second, the data on payments are only available at the municipal level and do not identify which specific *barangays* received the incentive payments.[3]

BACKGROUND, DESIGN AND IMPLEMENTATION

Background

The design of the NFB scheme coincided with the passing of the *National Water Crisis Act* as a vehicle to strengthen watershed management in the Philippines in the context of decentralization. The Department of Environment and Natural Resources (DENR) central office encouraged field offices to develop innovative initiatives to solve local forest management issues (Table 6.1). A DENR field office in Mountain Province (referred to in this chapter as the Community Environment and Natural Resources Office in Sabangan or CENRO-Sabangan) responded to this challenge with the concept of a no-fire bonus scheme.

Fire is a key issue in the Cordillera's pine forests of Mountain Province. A forest fire management project in the late 1980s determined that 99 per cent of forest fires in the Cordillera were caused by human activities, such as slash-and-burn agriculture, debris and garbage burning, arson and other indiscriminate use of fire (FAO 1989 cited in Pogeyed 1998). According to dendro-chronological analyses (fire scars in tree stems) and reports of upland villagers on large forest fire events, fires could burn whole villages (Pogeyed 1998). In 1993, a strong El Niño phenomenon contributed to the burning of large areas of pine forests (Table 6.1).

The timeline shows that the scheme ran for one monitoring and payment

Table 6.1   Timeline from scheme design to post-implementation

Date	Activity
1993–1994	El Niño: many uncontrolled forest fires in Mountain Province. Baguio *Midland Courier*, the local newspaper, featured the headline 'Mountain Province is burning'.
1995	CENRO-Sabangan prepared the No-Fire Bonus plan. CENRO proponents felt it was a 'crazy' idea, but because of the competitive spirit created by the DENR, they gave it a try.
1996	Scheme announced at the start of the year with support from Congressman and Governor, CENRO-Sabangan and local Department of Interior and Local Government helped inform *barangays*. CENRO-Sabangan forest rangers conducted two rounds of monitoring – first after the dry season (May–June 1996), then at year end (December 1996–January 1997), and observed a marked decrease in forest fire occurrences (based on anecdotal evidence, as baseline data on fires are no longer available).
1997	Certificates of no-fire occurrences were issued to 'winning' *barangays*, through municipal governments.
1997–1998	*Barangays* claimed their 'bonus' (incentive payment) in the form of infrastructure projects. CENRO-Sabangan asked the congressman for funds again, but none were available, as 1998 was an election year.
1997–1999	Incidence of fire remained lower than previous years. Villagers said that they were anticipating another round of rewards again. CENRO-Sabangan received an award from DENR and cited this scheme as one reason why they won. This boosted their morale and helped them to do their jobs better.
1998	The municipal officer of the Department of Interior and Local Government obtained an award and attributed this partly to her contribution to the contribution of the scheme to the Sagada Clean and Green Programme. International Forest Fire News featured the scheme.
2001	ASEAN Regional Haze Action Plan featured the scheme as a Philippines strategy.
2002–present	Some small-scale infrastructure projects visited during the fieldwork that had been selected by *barangays* as incentive payments under the scheme were still in good condition. Overall trend of fire occurrence maintained, secondary forest cover has increased and pine forest areas decreased. Total forest cover slightly increased.

cycle, but gained wide recognition and appeared to have a long-term social and environmental impact well beyond its funding cycle. We explore the potential reasons behind these observations later in the chapter.

The rationale for NFB needs to be understood in the context of the failure of traditional command-and-control strategies for fire protection in the Philippines. As Karki (2002, p. 7) notes, 'any analysis of forest fires needs to take into account the underlying causes of forest destruction.' An ancestral domain planning activity in one municipality in Mountain Province documented local people's analysis of the causes of uncontrolled fire, and the weakness of command-and-control measures in the area:

> Almost every year, forest fires occur despite ordinances prohibiting the burning of mountains and imposing fines on those proven to have started forest fires. The causes of *poo* [fire] include carelessness and sheer wickedness e.g. children playing with matches, passersby carelessly throwing cigarette butts on roadsides, and uncontrolled burning on farms, pasture lands, or areas for collecting honey' (Besao Ancestral Domain Management Plan 2002, p. 37).

The NFB scheme aimed to contribute to national watershed management objectives by decreasing the occurrence of uncontrolled fires, and contributing to rainwater infiltration, soil fertility, natural regeneration and enhancing wildlife habitat. Unlike previous national policies, which pursued forest conservation through command-and-control measures through the DENR, the NFB approach was premised on using financial incentives to secure watershed protection.

A distinguishing aspect of the scheme was the approach taken to delivering payments. Unlike the existing payment scheme of the state energy corporation, which was used to support an environmental fund for use in a quite narrowly delineated 'host community' where a hydro reservoir or a generating facility was located, the NFB scheme worked with communities across a large watershed area.

Apart from the limitations of command-and-control strategies, effective fire management has been plagued by other problems. The inadequacy of fire reporting and monitoring, as well as ambiguity over what constitutes a forest fire (Sutherland et al. 2004), makes it difficult to gain an accurate picture of the actual frequency and scale of forest fires. While reporting systems exist in the DENR to capture this level of information, several factors constrain field personnel in conducting this level of monitoring, the primary one being limited human resources at the field level.

Poor coordination between different line agencies is another key barrier to effective fire management. When DENR was reorganized in the late 1980s, the responsibility for forest fire management was transferred from its Forest Management Bureau to the Bureau of Fire Protection under the Department of Interior and Local Government. Although the Bureau of Fire Protection has the legal mandate for fire protection, its capacities are geared towards fires in built-up areas rather than forests. Meanwhile, the number of officers with forest fire management skills and experience in DENR had declined together with their budget allocations for forest fire management.

The NFB concept came out of the confluence of these factors. Informal discussions among field staff led to the proposal that was presented to local politicians. Since local politicians have funds for small projects in rural villages, they argued that these projects could be used as an incentive for fire prevention. Community participation in forest conservation and protection measures against fire outbreaks would be rewarded with community projects. In late 1995, the proposal gained the support of local political leaders (congressman and governor) who immediately agreed to the idea.[4]

DESIGN

The main similarity between the NFB scheme and the other PES schemes described in this volume is the common emphasis on financial incentives for providing an environmental service (in this case, watershed services). In every other respect, the scheme was 'PES-like' rather than an ideal PES scheme according to the criteria defined by Wunder (2005). (See Table 6.2 for a comparison of the NFB against Wunder's five criteria for PES.)

Environmental Service

As noted in Table 6.2, the NFB scheme aimed to secure watershed services through fire prevention, and through this maintain forest cover and natural regeneration. Watershed protection is a priority in the Cordillera Administrative Region, with 13 major watersheds with a drainage area of 21 101 km^2 and groundwater storage of about 75 305 million m^3 (NEDA 1994 cited in ESSC 2004). The region's water resources have the potential to supply energy and irrigation to meet the demands of its 1.3 million inhabitants and the rest of northern Luzon. Due to its watershed importance, the Cordillera Administrative Region is considered the 'watershed cradle' of northern Luzon. Eighty-one per cent of Cordillera Administrative Region's land area has been classified as public forest-lands[5] (DENR 2001a) and 6 per cent officially categorized as watershed forest reserves.[6] Six of the 125 watershed forest reserves in the Philippines

Table 6.2   PES criteria versus No-Fire Bonus scheme

PES criteria (Wunder 2005)	No-Fire Bonus scheme
1. A voluntary transaction where . . .	NFB was voluntary in that the reward was provided for voluntary measures by communities beyond what was legally required.
2. a well-defined environmental service (ES), or land use likely to secure that service	The environmental service (watershed protection) was not explicitly defined and measured. Fire avoidance was used as a proxy for determining if watershed services were delivered (that is, water table recharge, better infiltration and sustained water flow, maintenance of soil fertility, improvement in biodiversity). Direct monitoring of the environmental service did not occur; instead the scheme assumed that fire prevention would support watershed protection by maintaining forest cover maintenance and natural regeneration.
3. is bought by an (minimum one) ES buyer	Unlike PES schemes where the beneficiaries of the ES are the buyers, the NFB was paid for by a congressional representative from the target area.
4. from an (minimum one) ES provider	The ES providers were loosely dealt with at the *barangay* (rather than individual or household) level, with linkage to traditional community level institutions.
5. if, and only if, the ES provider secures ES provision (conditionality)	There was an element of conditionality as the incentive payment ('bonus') in the form of infrastructure projects could only be claimed on issue of a 'certificate of no-fire occurrence' to the relevant *barangay*. However, individuals and households belonging to the community could benefit from the incentive scheme (for example, by using the infrastructure and services it funded) even if they did not help in fire suppression.

are in Cordillera (DENR 2001b). The NFB scheme was implemented in municipalities in the headwaters and tributaries of the Abra, Agno, Magat and Chico Rivers. As such, the scheme was sub-national, and covered eight municipalities in Mountain Province under the responsibility of CENRO-Sabangan. It covered over 75 000 ha of forests in Mountain Province, of which around 30 per cent was pine forests.

Biophysically, it covered the conifer forests within the Central Cordillera Mountain Range which are the most fire-prone in the Luzon montane forest eco-region. The pronounced dry periods (November–April) and periodic fires favour natural regeneration of the Benguet pine (*Pinus insularis*) occurring at elevations between 1000 and 2500 m above sea level. The opening up of forest by human-set fires allowed the winged seeds of pine to penetrate into mossy forest areas highly valued for watershed services. Occasional burning enhances the natural regeneration of pine, but perennial burning can destroy pine saplings and even mature trees.

'Buyers', 'Sellers' and Intermediaries

As noted earlier, for the purpose of analysis in this chapter, the congressional representative and governor are regarded as the 'buyers', as they provided funding, while the *barangay* governments are considered 'sellers'. Other government line agencies served as intermediaries in establishing the scheme and facilitating the flow of benefits (Table 6.3).

As Table 6.3 shows, the scheme was not market-based as the ES buyers (local politicians) were using budget allocations from the national government. Nevertheless, the mechanism for delivery of the payment is of interest in the design of REDD schemes.

The ES providers (sellers) encompass two levels. The first was the *barangay* government which was the unit that received certification, facilitated agreement on the benefit package, received the payments and generated cooperation from households and individuals in forest fire prevention activities. The second level was the households and individuals who participated in different activities under the scheme, including forest fire prevention, suppression and construction of the infrastructure project awarded as a result of service delivery.

Table 6.3 indicates that the scheme had three intermediaries. The primary intermediary was the CENRO's Forest Protection Unit. This group was appointed as the ES monitoring body because forest fire investigation was a regular, albeit under-resourced, function of its forest rangers. A committee to identify which *barangays* would receive the NFB reward involved staff from the CENRO and the Environment and Natural Resources Office of the provincial government. The second intermediary was the Department of Public Works and Highways (DPWH), which served as the disbursing agent for the congressman. The third intermediary group comprised municipal governments that provided the environmental services sellers (that is, *barangay* governments) with logistical support and technical assistance in following up payment claims.

Table 6.3   Buyers, sellers and intermediaries

Actors	Role	Responsibilities
Congressional Representative of Mountain Province	Buyer	Allocated PHP100 000 (US$2223)* worth of infrastructure projects to each *barangay* without a forest fire during 1996. Funds came from the Congressman's Countryside Development Fund where he had full authority on disbursements.
Governor of Mountain Province	Buyer	Committed to provide matching funds of P100 000 (US$ 2223) for *barangays* without forest fire during 1996 (but failed to deliver on the commitment).
Barangay governments	Seller	Encouraged community members to cooperate in preventing forest fires and to volunteer in suppressing unwanted forest fires; Reported forest fires to the forest protection officer covering their *barangay*; Claimed the 'certificate of no-fire occurrence' from CENRO; Identified the infrastructure project to be proposed as incentive payment based on the *barangay* development plan; Bore the transaction costs of claiming the award.
Community Environment and Natural Resources Office in Sabangan Municipality (DENR field office)	Intermediary (ES monitoring body)	Monitored occurrence of forest fires through its Forest Protection Unit; Issued 'certificates of no-fire occurrence' to deserving *barangays*.
Municipal governments	Intermediary (administrative support)	Assisted *barangays* in following up with DENR, Congressman and Department of Public Works and Highways for implementation of agreed projects.
Department of Public Works and Highways in Mountain Province (provincial office)	Intermediary (disbursing agent of buyer)	Released funds for infrastructure projects to *barangays* based on endorsement of Congressional Representative and DENR's 'certificate of no-fire occurrence'.

Notes: * US$ 1 = PHP 45.

Incentive Payment

The payment was in the form of an infrastructure project to those *barangays* awarded a 'no fire certificate', rather than direct payments to individuals or households. These awarded *barangays* were to be provided PHP200000 (US$4444) worth of infrastructure projects. As such, the benefit delivery was performance-based. The actual budget accessed, however, was half of what was originally committed.

The programme aimed to revive traditional practices of forest fire management. Traditionally, community sanctions on perpetrators of uncontrolled fires were stringent. An offering of one pig was the penalty paid to the *dap-ay* (indigenous sociocultural institution composed of community elders) for starting a fire if it affected sacred grounds. In some communities, sanctions were levied not only on people who set fire to forests but also on people who happened to be in the vicinity during the onset of the fire and did not help in extinguishing the fire. The NFB scheme attempted to rejuvenate traditional practices by devolving the frontline responsibility for fire prevention, detection and suppression to *barangays*, a tier of government which is closely linked to traditional village institutions. Maintaining payments at the community level also aimed to foster community cohesion, unlike earlier 'bounty-style' schemes which had fostered competition and conflict as neighbours turned on each other to obtain a bounty for turning in those implicated in the spreading of fires.

Voluntary, Negotiated Framework

Like conventional PES schemes, the NFB scheme was voluntary and negotiated, since much of the land area covered by the scheme fell under ancestral domain. The scheme was founded on the consent of the indigenous groups residing in the area, although in formal terms the target area was classified as public forestlands.

Terms of Agreement

CENRO-Sabangan, which drafted the project concept in 1995 (which outlined the roles of the parties involved), presented the concept to the congressman of Mountain Province and to the provincial governor. Both elected officials agreed to allocate funds for the incentive payments, and thus became the 'environmental service buyers'.

The agreement between the CENRO and local politicians was oral. No legal agreement was drawn up to formalize the scheme: CENRO admitted that it was an oversight on its part and ultimately contributed to the scheme's non-sustainability. Since the politicians were eager for immediate implementation, CENRO's focus turned immediately to implementation and this important issue of a legal agreement was overlooked. In terms of the transaction with environmental service providers, the 'certificate of no-fire occurrence' became the de facto contract between *barangay* governments and the local politicians, since it served as proof of service delivery and stipulated the incentive payment due.

Conditionality

The scheme was conditional at the *barangay* level; payment was made to the *barangay* upon service delivery (that is, fire prevention/suppression) as evidenced by a 'certificate of no-fire occurrence' from CENRO. Only *barangays* that did not incur any forest fire during the dry season, or else made efforts to suppress its spread, received a certificate. Decisions on the choice of infrastructure projects rested upon each *barangay* government, which selected them from projects identified in the *barangay* development plan.

At the household level, the incentive to cooperate in fire prevention lay in the community benefits that certification for the *barangay* could bring in the form of infrastructure. At the same time, households that did not actively contribute to fire suppression could also benefit. According to the *barangay* officials who were interviewed, individuals could also gain cash income from participating during infrastructure construction as long as they were residents and had the skills needed. They no longer distinguished between members who cooperated in fire prevention and fire suppression where fire occurred.

Additionality

The NFB scheme exhibited additionality[7] due to the particular ecological conditions in Cordillera. In the area that the scheme covered, the occurrence of uncontrolled fires has been, and remains, a real threat. Fires in pine forests threaten the adjacent mossy forests, most valued for water regulation services (Walpole 2002). Annual burning has been eating up the edges of the mossy forest, thereby reducing its effectiveness in providing watershed services. Additionality has also been established because villagers of Cordillera, of which 92 per cent are from indigenous groups, mainly depend on upland farming and thus bear costs in the form of participating in patrolling activities and in forgoing the use of fire as an efficient tool for clearing upland farms and establishing pasture areas.

IMPLEMENTATION

The NFB scheme was implemented for one cycle covering a span of 3 years, from 1996 to 1998. The first year (1996) was the monitoring period for ascertaining which *barangays* fulfilled the scheme's criteria for certification under the NFB scheme (that is, did not incur fires). The second year (1997) saw the issuing of the 'certificate of no-fire occurrence' on which payment was based. The third year was when incentive payments were disbursed (Figure 6.1). Since the cycle coincided with the term of office of elected local government officials, the scheme terminated when the supportive officials ended their term because the scheme was not institutionalized in local policy.

Forest rangers conducted forest fire occurrence investigations after the dry season (May to June 1996), then evaluated again at the year end (December 1996 to January 1997). At the end of the investigation, they prepared a monitoring report, which became the basis for issuing the 'certificate of no-fire occurrence'. The certificate was given in 1997 to *barangays* to claim a one-time payment. However, when *barangays*

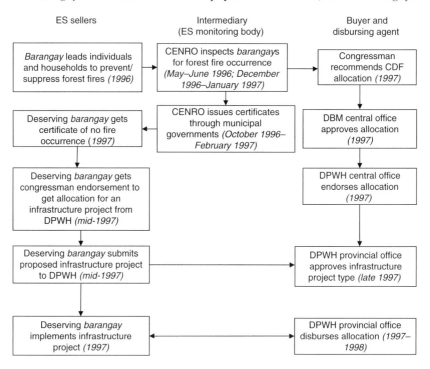

Figure 6.1   Process for claiming the incentive payment

started to claim their awards in mid-1997, only the congressman delivered on the funding commitment, as the provincial government was not able to provide matching funds. The congressman allocated around PHP5.2 million (US$115 555) from the Countryside Development Fund over which he had full authority. The provincial governor, on the other hand, was not able to deliver on his commitment, as he was operating under a more stringent budgeting system and differently timed budget cycle, and so was not able to tap regular provincial funds for the scheme.

Participation

Three aspects of participation are analysed in terms of providing the environmental service and in accessing benefits. These include the eligibility to participate, willingness to participate and ability to engage with the scheme (Table 6.4).

From the buyers' perspective, both of whom operated at the provincial level, the eligible participants were the 124 *barangay*s with fire-prone pine forests located in the eight municipalities covered by the CENRO-based Sabangan Municipality.[8]

In terms of service delivery, the willing and able participants (participation in practice) were the 121 *barangay*s awarded with a 'certificate of no-fire bonus occurrence'. Within these *barangay*s, only some households actively helped to prevent, detect and suppress forest fires when the scheme was in effect. These *barangay*s became eligible to claim the incentive payment.

In terms of benefit delivery from the incentive payment, this went to *barangay*s who implemented activities for fire protection, which is estimated to be only around 43 per cent or 52 *barangay*s. Data are not available as to why some *barangay*s did not claim payments. Indications are that transaction costs may have been a factor (for example, transport costs to follow up the payment) but further investigation is needed to gain an understanding of this important issue, which could ultimately become a disincentive to act.

Within *barangay*s that claimed incentive payments, benefit delivery occurred in two ways. The first way involved residents with the required skills who were hired as workers during construction of the chosen infrastructure projects, thus obtaining cash income. The second involved residents that benefited from the infrastructure project. Based on interviews in seven *barangay*s, around 5–15 per cent of households had members hired as workers during infrastructure construction and between 50 and 100 per cent of households benefited from the completed project, depending on the type and location of the infrastructure. For instance, in Banguitan,

Table 6.4   Eligibility, willingness and ability to participate

Form and scale of participation	Eligibility	Participation in practice
Service delivery		
Barangay	All *barangays* located in Mountain Province with pine forests within jurisdiction of CENRO-Sabangan (124 *barangays*)	Majority of *barangay* leaders successful in coordinating fire prevention/ suppression actions (121 *barangays* = 97% of eligible *barangays*).
Household	All residents of eligible *barangays*	Households in the 121 *barangays* helped in: • prevention ('passive' but primary participation by not starting a fire or managing fires properly); • detection (spreading information on fires); • suppression (putting out uncontrolled fires) (figures not known).
Benefit delivery		
Barangay	All *barangays* gaining 'certificates of no-fire occurrence'	Leaders willing to devote time and personal funds in going through the process of claiming the incentive payment. In the case of Besao municipality, only six villages (43%) claimed the award, perhaps due to their ability to cover transaction costs of claiming the incentive (infrastructure project). Two villages (14%) did not claim due to the high transaction costs and the rest had no recall of having claimed the payment.
Household	Residents of *barangays* with 'certificates of no-fire occurrence'	Residents of *barangays* willing to go through the process of claiming the incentive payment and having the skills needed in the selected infrastructure project (around 10–15% of households) earned cash income. Residents benefited from completed infrastructure (50–100% of households in a *barangay* depending on type of infrastructure).

men from ten households (around 8 per cent of total households in the *barangay*) were employed to construct the erosion-control stone wall for 2 weeks. The stone wall structure benefited 100 per cent of the households because it protected the only road access to the *barangay*.

Monitoring

CENRO fire investigation reports were provided to a multi-stakeholder committee (discussed earlier) to determine which *barangays* could be certified under the programme. These data were also part of a regular quarterly reporting system through the DENR provincial and regional office. The regional office was responsible for consolidating fire reports annually and for providing historical comparisons.

Fire investigation reports contained information which included the date of fire occurrence, the exact location of fire occurrence and the number of hectares affected, a time record (time observed, time message received, time of crew dispatch, time travelled from base to fire site, time fire controlled, time fire put out), the character of the fire, and the nature and cost of damage.

In practice, however, limitations in human resources and logistical support hindered thorough monitoring and provision of technical support. When the scheme was implemented, the Forest Protection Unit had 20 forest rangers assigned by sector, covering one municipality per sector. Each sector was assigned between one and six rangers, depending on the characteristics of the municipality covered. Each ranger covered an average of 3750 ha, with limited time and resources available to fully investigate each *barangay*, leading to a less than reliable database on fire incidence.

If a fire occurred along boundaries of neighbouring *barangays*, then neither *barangay* would be eligible for the incentive. However, implementing this was a challenge. A *barangay* official raised a situation for consideration, related to the fairness of this sanction: if a fire was started in an adjoining *barangay* and then spread to a second area but the second made an effort to suppress the fire moving toward their jurisdiction, should they still be considered ineligible for the incentive? Although fire investigation reports were supposed to document the perpetrator, all reports that were reviewed indicated that the perpetrator was 'unknown'.

Limited follow-up monitoring was done after the certificates were issued and files turned over to the congressman's office to facilitate the award of projects. It was up to the *barangays* to show their certificate to the congressman and claim their award from the DPWH.

Transaction Costs

The transaction costs that *barangays* incurred to prevent or suppress forest fires included time spent for community consultations and meetings to raise awareness and to agree on how to coordinate efforts when fires started. No direct monetary compensation was provided to individuals who participated in fire suppression. It was an activity to which people, being part of the community, were traditionally expected to contribute. During the focus group discussion in *barangay* Banguitan, 11 traditional leaders described their way of coordinating putting out forest fires. While it was the generally the men, including young boys, who were involved in the actual putting out of fires, women and children would also help in spreading word of a forest fire. Fires were put out by hitting burning branches or the ground and embers with tree branches, making a fire line, carrying water, or using hoses connected to rice fields if there was water there.

Some *barangay* officials used personal funds to meet costs incurred in following up with CENRO the release of their no-fire certificates. The greatest costs were incurred, however, during the process of claiming their reward from the congressman, the governor and the DPWH.

The process of getting the congressional funds released for the infrastructure projects was very complex. According to the DPWH, the process started with the congressman giving instructions to the central office of the Department of Budget and Management which then transferred funds to DPWH central office, which in turn remitted the funds to DPWH provincial offices.

Payments for infrastructure projects were provided on a reimbursement basis. Some *barangays* received less than this amount due to deductions made by DPWH for training expenses. The DPWH, responsible for disbursing the funds under the scheme, was supposed to provide training to *barangay*s to aid their implementation of the awarded infrastructure project. One *barangay* related that it only received PHP80000 (US$1777) because the DPWH deducted PHP20000 (US$444) for training expenses that DPWH claimed had been incurred. The *barangay* official reported, however, that no training was actually conducted in their area, indicating a misuse of funds.

While risk in a PES scheme is often seen from the perspective of the environmental service buyer, the NFB scheme illustrates that environmental service sellers also face risks, particularly where a scheme is not underpinned by a clear legal agreement. First, one of the environmental service buyers (provincial government) defaulted on its payment, which reduced the promised incentive payment by 50 per cent. *Barangays* then had to rework their proposed projects to fit within this reduced amount,

---

BOX 6.1 TRANSACTION COSTS IN *BARANGAY* BANGUITAN, MOUNTAIN PROVINCE

Barangay officials informed the community that they should do their part to ensure that there would be no burning in the forests. When a forest fire was detected within the *barangay*, the officials led other residents in putting it out. After the dry season of 1996, a 'certificate of no-fire occurrence' was issued by CENRO for all 14 *barangay*s in Besao Municipality, including the *barangay* of Banguitan.

To obtain the reward, Banguitan needed to undertake the following activities, the monetary values of which are not known:

- a *barangay*-wide meeting to identify the priority infrastructure needed, based on the *barangay* development plan. The proposal was for a drainage canal and trail pavement because the present one was unsafe to walk on during the rainy season, especially for children.
- meetings to draft and pass a resolution to support processing of the reward.
- several trips to Bontoc (3 hours away by sparse public transport) to process the drainage canal proposal with the congressman and the DPWH.

Additional trips had to be made after Banguitan was informed by the DPWH that the available funds were inadequate for a drainage canal and they had to choose another project. Because of this, Banguitan had to repeat the three activities detailed above. Available funds could only support an erosion-control stone wall of a roadside that was very erodible. When the congressman approved this second proposal, the *barangay* captain then went to the DPWH, but found the process to be difficult (Figure 6.1). Banguitan and other *barangay*s in Besao sought support from the municipal government when following up with the DPWH.

---

creating additional transaction costs in an already complex process (Box 6.1). While no contractual obligations were technically broken and no legal sanctions were imposed on the environmental service buyer, the governor was not re-elected, which may have been related to social sanctions related to the strong social networks in Mountain Province.

No additional transaction costs were incurred by the intermediaries. Processing infrastructure development contracts was part of the regular functions of the DPWH and fire investigation was a regular part of the responsibility of the Forest Protection Unit of the CENRO. Personnel from other units such as reforestation, watershed, and Community-Based Forest Management were tapped to assist in fire monitoring, suppression and investigation operations. However, CENRO had only a very small budget for funds for field work; they therefore incurred opportunity costs of field work for alternative purposes.

IMPACTS AND CHANGES

Impacts on Livelihoods

The most immediate and direct impact of the NFB was the financial benefit obtained by households that provided labour, materials and services during the construction of the infrastructure project selected. Individuals who helped accomplish the documentation requirements and financed travel and other costs of processing the papers with the four government agencies (DENR, Congressman's Office, Governor's Office, DPWH), may have also gained some income.

In Banguitan,[9] ten men were employed to construct the erosion-control stone wall for 2 weeks. Construction was timed to occur during the lean season of the agricultural cycle, hence no opportunity costs of forgone farming income were incurred in this *barangay*. Each worker was estimated to have received PHP116/day[10] (US$2.50) or PHP1624 over 2 weeks (US$36). Using these assumptions, the scheme provided additional wage income of PHP16240 (US$360) to ten households in the *barangay*. Although short term, this was considered a major contribution, since the average income of families in Banguitan before the scheme was PHP1598 per month (US$35) (Besao Municipal Government 1991).

The scheme achieved 97 per cent participation among the pool of environmental services sellers from the *barangays*. Table 6.5 shows that of the 124 *barangays* covered by the scheme, 121 *barangays* were granted certificates of no-fire occurrence. However, only 52 *barangays* (or 43 per cent) were estimated to have received the incentive payment. One reason cited for not following up on the award was the high transaction costs of processing the claim. Aggregate wages could have varied per project depending on the number of men hired and number of person-days needed to complete the projects.

The scheme also generated positive distributional impacts[11] on *barangay*-

Table 6.5   Villages certified with no-fire occurrences, 1996

Municipality	Total villages	Villages certified	Villages not certified	Villages awarded (est.)
Barlig	11	11	0	5
Bauko	22	22	0	9
Besao	14	14	0	6
Bontoc	16	16	0	7
Sabangan	15	14	1	6
Sadanga	8	8	0	3
Sagada	19	17	2	7
Tadian	19	19	0	8
Total	124	121	3	51

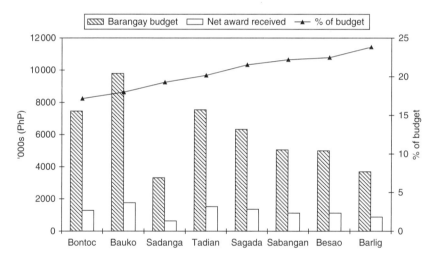

Figure 6.2   Impact on barangay *financial capital, 1998*

level financial capital in 1998 (Figure 6.2), apart from direct household benefits to those it employed. The flat-rate incentive payment had a more pronounced impact on *barangays* with relatively lower budgets from national governments. This differential impact on low-budget *barangays* can be seen in Figure 6.2. The scheme increased the financial capital of Barlig and Sadanga *barangays*, those with the lowest annual budgets, for 1998 by 24 per cent and 19 per cent, respectively. In contrast, Bauko and Tadian *barangays*, with the highest budgets, obtained a relatively lower proportional increase in their financial capital (18 per cent and 20 per cent respectively).

Table 6.6   Direct community benefits of awarded projects, Besao Municipality

Awarded project	Direct benefit	Impact on community
Erosion control	Road protection	Better access to hospitals, schools, markets (for example, Banguitan village)
Drainage canal	Improved water flow	Decrease in water-related diseases (for example, Besao West village)
Water tank	Improved water access	Availability of water during dry season (for example, Gueday village)

Other direct benefits from the infrastructure projects in Besao Municipality are summarized in Table 6.6. These infrastructure projects are still operational and in use 10 years after construction. The sense of ownership of the project – gained in large part from their effort to control fire and which they themselves had selected – led to better maintenance of the infrastructure.

A number of longer-term and more broadly shared impacts were derived from the NFB, including:[12]

- providing parents with a venue to reinforce in the youth the cultural value of participating in forest fire prevention and suppression;
- encouraging elders to pass on to the youth traditional fire suppression methods;
- giving *barangay* leaders a chance to practice leadership skills among their constituents, and negotiation skills with provincial governments and line agencies;
- creating sociocultural events for people to gather and work together towards a common goal, for example, during fire suppression and construction work;
- providing a venue for building relations between local governments and communities;
- inspiring some municipal governments to replicate some aspects of the scheme and apply their own innovations using available budgets.

Impacts on non-participants were investigated by inquiring among *barangay* leaders about possible cases of people who were negatively affected by fire control or by the construction of the awarded infrastructure projects. Respondents enumerated reasons why and when fire is beneficial to livelihoods, which provides insights on the negative impacts of fire suppression:

- Fire induces growth of fresh grass for cattle. Some cattle raisers think that pine forests and grazing are incompatible land uses because pine needles hinder grass regeneration. In Kinali *barangays*, burning in communal pasture areas is allowed when it is likely to rain so that the fire can be contained. On the other hand, other cattle raisers discourage burning as a means to induce grass growth as frequent burning destroys soil quality. If the grazing area is located on steep terrain, new forage growth invites cattle to go to these steep areas, which could cause them to fall and die.
- A newly burned mountain slope is good for growing legumes. This is rarely practiced now, but if there is a newly burned slope, any community member may be allowed to cultivate legumes in the area over one cropping period, with the understanding that the land still belongs to the owner.
- Fire helps control rats and other pests, benefiting sweet potato and rice fields near the area burned. However, while this can be beneficial to some, it can be disadvantageous to the owners of rice fields and croplands that abut burned land.

In terms of possible inter-municipal tensions due to the limited coverage of the scheme, the scheme proponents in CENRO reported that such tensions did not occur. The scheme was understood as a pilot programme run and initiated by CENRO-Sabangan and applied only to areas where pine forests predominated. However, proponents also said that the probability of tensions emerging could not be discounted if the scheme had been implemented for a longer period without being inclusive of municipalities in the other forest district within Mountain Province.

CENRO personnel who were involved in the scheme's implementation noted that they were aware of the possibility that a fire may start in a *barangay* along the border with another *barangay* if tensions were present. The scheme highlighted the need for neighbouring *barangays* to clarify how responsibilities should to be shared in suppressing forest fires occurring along a shared boundary. While they did not encounter any such case in 1996, they warned not to discount the possibility of tensions arising between neighbouring *barangays* if the scheme was implemented for a longer period, the selection and awarding process became more politicized, and the reward was increased significantly.

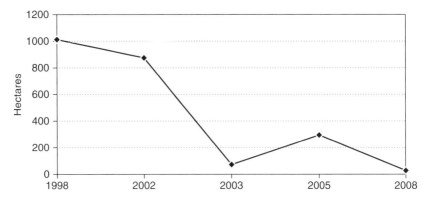

Figure 6.3   Forests affected by fire (total hectares), Mountain Province

ENVIRONMENTAL CHANGES

The objective of the NFB scheme was protection of forests against uncontrolled forest fires. Municipal-level baseline statistics before the scheme's implementation were not available to allow comparison with the post-implementation trend (Figure 6.3), which shows a declining trend in the extent of forest fires after one cycle of implementation. In Mountain Province, available data indicate that forest cover has increased by about 1 per cent over 15 years, particularly mossy and secondary forests, while pine forests have decreased. The changing forest composition may have a linkage with fire reduction as pine cannot naturally regenerate without fire, giving broadleaf species a chance to survive. However further validation is needed to attribute such changes to the NFB scheme.

LESSONS FOR REDUCED EMISSIONS FROM DEFORESTATION AND FOREST DEGRADATION (REDD)

Factors Contributing to Viability and Livelihood Impacts

This scheme became feasible in this context because of the interplay of the following cultural and sociopolitical factors.

1. *Culture.* The indigenous peoples of Mountain Province, like other indigenous groups in the Cordillera Administrative Region, hold a strong sense of ownership and responsibility for their lands and

forests, and practise cultural systems of resource management handed down across generations. Even though 81 per cent of the land in Cordillera Administrative Region has been nationalized, pre-existing community agreements govern use allocations of forests and wood-lots to each family/clan, and which forests are considered to be under communal ownership. This strong sense of ancestral domain owner-ship and responsibility, combined with values of cooperation, volun-teerism and reciprocity, facilitated the cultural norm of cooperating when a forest fire needed to be suppressed. In recent times, however, elders have observed that these values are being eroded. The scheme was appreciated because it gave the communities an opportunity to rekindle these values.

This has been supported by the fact that the NFB scheme built upon existing community institutions and involved traditional leader-ship. It showed that the *barangay* government, with support from the municipality, was an effective way to 'bundle' smallholders, particu-larly in cases where village governance was tied to cultural systems of leadership.

2. *Tenure and resource access*. Experiences of marginalization and injus-tice created the tendency for indigenous communities to resent government-imposed laws. One such law was the national forest policy which had disfranchised them from their ancestral lands. This prompted the attitude 'Why protect the forest when government does not acknowledge that it belongs to us?' The new tenure instruments in 1996, such as certificates of ancestral domain claims and community-based forest management agreements, formally recognized the indigenous resource management arrangements described above.

3. *Scale*. Designing and managing the scheme at sub-national levels (provincial and municipal) enabled the scheme to more efficiently reach the village level – the best level to prevent, monitor and suppress forest fires before they become unmanageable. Although short-lived, the NFB scheme was starting to build on local leadership, linking incentives to local resource management systems and promoting multi-sectoral partnerships, with the scope to generate more socially optimal results than previous fire protection efforts.

4. *Monitoring system*. The NFB scheme used fire incidence as a proxy for watershed services, which was already backed by an existing, albeit weak, monitoring mechanism within the CENRO. The monitoring system did not distinguish whether a forest fire caused deforestation or degradation, and rewarded forest fire detection and suppression as well as prevention. Investigators applied sanctions only to villages where reckless and uncontrolled burning occurred, and not when

people were using fires responsibly as tools for resource management. This helped the scheme to accommodate traditional agricultural practices, which would ultimately reduce adverse impacts on local livelihood activities.

Monitoring is needed not only for the impact of incentive payments on environmental services delivered but also for the effectiveness of incentive payment delivery. In the NFB scheme, the CENRO failed to monitor which certified no-fire *barangay*s actually received payments from the DPWH and so was not able to help communities solve problems in claiming the award. This is important to maintaining the trust of *barangay*s in the scheme. Otherwise, unresolved issues can become disincentives to act.

5. *Reference scenario.* Analysis of stakeholders' perceptions of 'uncontrolled' fire is a very crucial step. The baseline scenario for the NFB scheme was based on region-wide historical data that enabled the impacts of the scheme on fire incidence to be established. The underlying drivers of uncontrolled forest fires were identified by district-level officials who were from the target communities and, therefore, had a sense of what incentives might be attractive to them.

6. *Mode of benefit delivery.* The NFB scheme shows that in areas where cultural cohesion is strong, an incentive scheme that provides benefits not only to individuals (for example, income during construction) but also to the entire community (for example, the completed infrastructure) may be an effective approach. While the money invested in infrastructure works was small, the framework and the financial mechanism were closely linked to local environmental and social objectives and existing community institutions.

7. *Payment timing and conditions.* The incentive payment was conditional on delivery of the service. This eliminated the risks associated with up-front payments and used existing monitoring systems to reduce monitoring costs.

8. *Intermediary role of local governments and other line agencies.* The NFB scheme was resourceful in using existing systems within local governments and other line agencies outside the DENR. Transaction and opportunity costs to smallholder environmental services sellers may be reduced through such an approach, which taps existing systems in government and supports knowledge sharing to improve the negotiating position of environmental services sellers with environmental services buyers. In the NFB case, the negotiating position of *barangay* environmental services providers improved through the support of municipal governments.

Factors that Undermined Sustainability and Broader Effectiveness

While the scheme may have had many positive elements, major flaws in design and implementation ultimately would have undermined its sustainability and its effectiveness if it was implemented over a longer period or a broader area. These factors include the following.

1. *Source of funds.* Publicly funded schemes such as NFB can be held captive to short-term political agendas (Wunder 2005). For this reason, some scholars do not regard local politicians as legitimate 'buyers' of environmental services. While they may have the funds, they do not really have a compelling need to secure the services over time.

   Tapping the commitment, goodwill and buy-in of politicians enabled the scheme to proceed at a pilot level, but also brought costs. First, reliance on politically motivated sources of funding subjected the scheme to agendas that were seen to jeopardize its objectivity. Since the award period occurred during an election year, the scheme became politicized. This created difficulties for some villages (that is, those that were not allied with the congressman) when claiming the incentive payment, which showed that willingness-to-pay was really motivated by expected 'political returns' rather than environmental service delivery.

   Furthermore, since the agreement between the CENRO and the local politicians was oral and not translated into policy, the scheme became vulnerable to changes in political leadership and priorities. Priorities shifted after the elections, leading to the discontinuation of funding for the scheme.

   Apart from the possibility of funding the scheme through market mechanisms, the NFB scheme would have had a better chance at sustainability as a government initiative if:
   - its funding was integrated into regular programmes of government and explicitly linked to the *Water Crisis Act* and other policies that established funds collected from downstream users to support water resource regeneration;
   - a written agreement was drawn up, signed and translated into policy with a regular programme budget.

2. *Monitoring capacity and data availability.* The impact analysis assumed that certified *barangays* did not have any uncontrolled forest fires when the scheme was in effect. However, it is also possible that forest fires occurred but that these were not reported and the forest rangers in the investigation committee were not able to independently verify the occurrence of forest fires, particularly in areas not visible from settlements or roads.

Monitoring, reporting, verification and compliance could be improved if (a) adequate operational funds were provided to district forest offices and local governments, (b) local capacity in geographic information systems analysis and accounting for environmental services (including carbon) was built, and (c) coordination opportunities between local government institutions were made available.

3. *Climate variability.* Two other related factors contributed to curbing forest fires in 1996: forest type and climate. DENR observations in the field suggest that severe fires occur in pine forests approximately every 5 years, due to the available fuel loads. The widespread burning during the El Niño of 1993–94 meant that pine forests in Mountain Province had low fuel loads in 1996, making them less prone to severe and extensive fires. However, more generally, the increasing frequency of El Niño episodes has increased fire risk (probability of occurrence), with increases in areas affected by fire in 1998 and 2005 coinciding with El Niño periods.

4. *Disbursement system and transaction costs.* The scheme provided a preview of the complexities that need to be anticipated if a REDD scheme is to be managed across different government agencies. Even though the scheme was innovative in expanding the involvement of line agencies and resourceful in using existing systems for disbursing the payment through the DPWH, high transaction costs deterred several fire-preventing *barangay*s from claiming their award. The disbursement system should have been communicated to participants upon announcement of the scheme. Participants should be informed in advance of any anticipated reductions in the payments committed.

5. *Social marketing.* The NFB scheme had been efficient in spreading news about the scheme to many *barangay*s through the local media, but it would have been more effective if greater efforts were made to explain clearly the linkage between the incentive payment and the desired environmental and livelihood impacts, and enable a better integration of the social dimensions of the scheme.

The key impacts are summarized in Table 6.7. Note that these observations particularly relate to the seven *barangays* that were visited during this study.

CONCLUSIONS

This study of the NFB scheme highlights the claim that, although it only operated for a short period, the scheme was starting to generate positive

Table 6.7   Summary of changes and impacts

Type of capital	Impact
Cultural	<ul><li>re-invigorated indigenous institutions and community forest fire management practices</li><li>nurtured existing cultural values of volunteerism and cooperation in caring for the environment</li></ul>
Physical	<ul><li>established new infrastructure in an estimated 52 villages. In the study areas, these show signs of continued community maintenance</li></ul>
Human	<ul><li>improved knowledge and skills in forest fire management, particularly amongst youth</li><li>improved awareness of health hazards brought about by forest fires amongst key informants</li><li>improved access to basic services in study areas with erosion-control infrastructures</li></ul>
Financial	<ul><li>increased *barangay* government budgets in 1998 by 18–24%</li><li>provided additional wage income to men who provided labour during infrastructure construction. Work occurred during the agricultural cycle's lean season, hence no opportunity costs were forgone</li><li>protected private and government economic investments from fire in participating *barangays* (crops, orchards, roads, waterworks)</li><li>reduced requirement for investment in reforestation due to natural regeneration</li></ul>
Social	<ul><li>provided communities with an opportunity to collaborate on a common objective (fire protection to gain community benefits)</li></ul>
Political	<ul><li>introduced environmental criteria for allocation of politicians' discretionary funds</li><li>resulted in greater connection of infrastructure funding with community priorities</li></ul>
Natural	Limited data, but:<ul><li>potentially contributed to declining trend in extent of forest fires over the last 12 years</li><li>potentially contributed to 1% increase in province forest cover, particularly of broad-leaf species, over the last 15 years</li></ul>

social and environmental changes during and beyond the implementation cycle. Key impacts are summarized in Table 6.7.

El Niño events are expected to become more frequent and perhaps more intense over time. The value of having a scheme such as the NFB in place is that the impacts of El Niño on fire can be mitigated. The 'selling' period could be timed to coincide with years that are projected to have increased forest fire incidents, for instance, in terms of El Niño event predictions, fuel load build-up and election years. Climate predictions would be needed so that the selling periods could be programmed over multiple years. This scale of event should also be considered in the design of incentive payment mechanisms under REDD.

Although the scheme contributed to generating positive livelihood impacts, several aspects of the scheme's design and implementation ultimately undermined its sustainability and broader effectiveness. While not a classic PES scheme, NFB does raise some important lessons for the design of future REDD schemes.

Notwithstanding its shortcomings, the NFB scheme treated forest fire management as a wider social objective, not a narrow technical problem. This enabled some useful, locally grounded activities to be supported through the scheme. It may be that a scheme like the NFB, with its emphasis on communal action and benefits, may be more viable in a context where national policies have limited reach and forest department human resources are limited, but there is strong basis for cultural cohesion. In the Cordillera, the strength of indigenous institutions and the political context seem to be critical factors. Existing collective institutions for environmental services sellers, such as community forest management federations and networks, tree farmer associations, indigenous peoples' networks and landscape alliances among local governments, already exist in the Philippines. These institutions may be tapped to serve as implementing institutions in the case of national REDD mechanisms and may serve as coordinating bodies for technical assistance to environmental services sellers.

Another important point was the existence of a context that enabled recognition of rights to land and traditional institutions for resource management. The joint policy of the DENR and the National Commission on Indigenous Peoples formally recognizing indigenous forest management systems and practices (2008) strengthened the existing ancestral domain legislation. Greater recognition of the resource management rights of indigenous forest owners provided further incentive to continue cultural systems of forest fire management.

As a sub-national scheme, the NFB was a product of a ripe policy environment, existing data collection processes, which could be easily harnessed for monitoring, and collaboration between scales of government. It highlights, however, that without a clear legal agreement, there is no basis for sustainability, and such an agreement needs to specify the rights and responsibilities of different actors, criteria for decision making (to avoid politicization), payments amounts and schedules and guidelines for obtaining payments.

On the final point of payment guidelines, although high transaction costs are often considered as part of establishing a scheme, they can be equally significant at the point of claiming the incentive payment. In the NFB case, it proved a significant deterrent to around half the *barangays* involved in the scheme.

ACKNOWLEDGEMENTS

We would like to thank the following individuals and organisations for their assistance in this research: Manuel Pogeyed, the Municipal Government of Besao, the people of Besao, the Mountain Province Environment and Natural Resources Office, the Sabangan Community Environment and Natural Resources Office, the Cordillera Administrative Regional Environment and Natural Resources Office, the Department of Public Works and Highways, Peter Walpole and the Asia Forest Network secretariat.

NOTES

1. Field visits were conducted by Rowena Soriaga of Asia Forest Network and Dallay Annawi of Environmental Science for Social Change. Ms Annawi, who grew up in Besao, Mountain Province, and who is a member of the indigenous group living in the case study site, arranged the visits and facilitated the meetings.
2. A *barangay* (village) is the smallest administrative unit of government. A municipality consists of several *barangays*, while a province consists of several municipalities.
3. At the time of writing, records on *barangays* that took up the award and the types of infrastructure projects awarded were no longer available in CENRO and DPWH. As government offices dispose of records every 5 years and 12 years have passed since the scheme's implementation, any files pertaining to the scheme have now been discarded.
4. Due to the decentralization of government authority, technical line agencies view local governments as institutions that can provide longer term, secure funding compared with development assistance sources. This stability in funding can only happen, however, if an initiative is institutionalized, that is, backed by a local ordinance and integrated in regular budgeting processes.
5. *Public forestlands* are the mass of lands of the public domain which have not been the subject of the present system of classification for the determination of which lands are needed for forest purposes and which are not (Presidential Decree 705, Section 3a).
6. Watershed forest reserves refer to those lands of the public domain which have been the

subject of the present system of classification and determined to be needed for watershed services (Presidential Decree 705, Section 3b).
7. Additionality is demonstrated when it can be proven that the service would not be secured to the same extent without the intervention in question.
8. The 19 *barangays* under the CENRO based in Paracelis, Mountain Province, were not included in the target participants because pine forests do not predominate there.
9. The population of Banguitan in 1990 was 618, of which 180 were working-age men (15–64 years old) (Besao Municipal Government, 1991).
10. Extrapolated using the 2008 rate of local wage employment of PHP210/day ($US4.60), discounted at 5 per cent.
11. Impact was estimated based on: (1) a claim rate of 43 per cent among certified *barangays*; (2) PHP80 000 net award per *barangay* (US777); and (3) 2007 internal revenue allotment data discounted at 5 per cent over ten years (DBM 2007).
12. Based on key informant interviews and focus group discussions.

REFERENCES

Besao Ancestral Domain Management Plan (ADMP) (2002 unpublished).
Besao Municipal Government (1991), Besao Municipal Profile, unpublished document prepared by Besao Municipal Government (Planning and Development Office).
Department of Budget and Management (DBM) (2007), Internal revenue allotment for *Barangays* FY 2007, http://www.dbm.gov.ph/index.php?pid=4&id=75 (accessed 19 October 2008).
Department of Environment and Natural Resources (DENR) (2001a), Land classification by region, http://www.denr.gov.ph/article/articleview/733/ (accessed 20 July 2004).
DENR (2001b), Watershed forest reserves by region, http://www.denr.gov.ph/article/articleview/737/ (accessed 20 July 2004).
Environmental Science for Social Change (ESSC) (2004), Strengthening community-driven natural resource management in the Cordillera Region, unpublished report submitted to AFN Community Forest Management Support Project in Southeast Asia.
Food and Agriculture Organization of the United Nations (FAO) (2006), Fire management: voluntary guidelines. Principles and strategic actions, Fire Management Working Paper 17, Rome: FAO.
FAO (2007), *State of the World's Forests*, Rome: FAO.
Karki, S. (2002), Community involvement in and management of forest fires in South East Asia, www.fao.org/forestry/media/11241/1/0/ (accessed 19 May 2008).
Pogeyed, M.L. (1998), No fire bonus plan program of mountain province, in International Forest Fire News, IFFN No. 18, January 1998, www.fire.uni-freiburg.de/iffn/country/rp/rp_4.htm (accessed 19 May 2008).
Sutherland, D., B. Arthur, R. Goze and S. Batcagan (2004), A review of forest fire management in the Philippines, ITTO, Yokohama, www.itto.int/direct/topics/topics_pdf_download/topics_id=6210000&no=1 (accessed 8 July 2008).
Walpole, P. (2002), An analysis of drivers and impacts of landuse change in the tropical uplands of Mindanao, Philippines, PhD thesis, King's College London.

Walpole, P., G. Braganza, J.B. Ong, G.J. Tengco and E. Wijangco (1993), Upland Philippine communities: guardians of the final forest frontiers, Research Network Report No. 4, Berkeley: University of California and Asia Forest Network & Centre for South East Asia Studies.

Walpole, P., G. Braganza, J. Ong and C. Vicente (1994), Upland Philippine communities: securing cultural and environmental stability', Research Report, June 1994, Manila: ESSC (formerly Environmental Research Division, Manila Observatory).

Wunder, S. (2005), Payments for environmental services: some nuts and bolts, CIFOR Occasional Paper No. 42, Indonesia: Centre for International Forestry Research.

7. Social and environmental footprints of carbon payments: a case study from Uganda

Laura A. German, Alice Ruhweza and Richard Mwesigwa with Charlotte Kalanzi

INTRODUCTION

This chapter presents a case study of the Trees for Global Benefits Programme in Bushenyi District, Uganda. The aim of the study is to describe the programme and the social and ecological impacts to which it has given rise. Findings from structured and semi-structured interviews with key informants, beneficiary and non-beneficiary households suggest that even with modest shifts in land-use patterns being induced by carbon offsets in the voluntary market, positive and negative social and environmental impacts can be significant.

Uganda's forests and forest products are vital in terms of their contribution to rural incomes and livelihoods. Seventy-five per cent of villages sell tree products, communities with access to woodlands benefit from a wide range of tree products and services, and 93 per cent of national energy consumption is from wood fuel and charcoal (MWE 2007). Recent estimates indicate that forest cover had declined from 24 per cent to about 15 per cent between 1990 and 2007, due to the pressure from population expansion and demand for fuelwood and timber (MWE 2007), industrial logging pressure (Welch Divine 2004) and armed conflict in northern Uganda. As a result, it is anticipated that there will be a shortfall in the supply of wood products, particularly timber, within the next ten years.

In 2002, Uganda carried out a Forestry Sector Review[1] aimed at aligning the forestry sector with the country's poverty reduction programme and sustainable development goals. The study identified a number of constraints to afforestation and the sustainable use of indigenous forests, including insecure and overlapping rights to land, the low market value for timber and other forest products, limited access to market information, poor road access and over-harvesting in natural forests.[2] The review also identified measures to address these constraints, including the establishment of incentive mechanisms, such as favourable taxation regimes for foreign investors, long-term land leases for afforestation on government land and permits for small-scale growers to grow trees in forest reserves. These measures also included a recommendation to access global financing mechanisms for carbon sequestration in forestry, such as the World Bank Carbon Fund, and carbon trading mechanisms provided for under the Clean Development Mechanism (CDM). The review also noted that, although it would be possible for Uganda to benefit from carbon offset projects, the absence of financial and technical guidelines hindered their ability to capture these benefits. The Uganda Forestry Sector Coordination Secretariat therefore commissioned a study to explore ways of establishing a functional forestry sector carbon project that would also enable the development of model financial and technical guidelines for enabling access to the carbon market. The terms of reference for the study included, among others, the following outputs:

- baseline data on carbon stocks for forestry, in order to enable subsequent determination of additionality;
- criteria and indicators for forestry projects based on international principles and national policies;
- designation of an institutional home for the approval of CDM projects;
- land for carbon offset projects made available and marketed;
- a Forestry Fund established in line with CDM modalities; and
- capacity built into national institutions (government institutions, non-government organizations and the private sector) to enable them to take advantage of carbon offset funding possibilities.

The Edinburgh Centre for Carbon Management (ECCM) was contracted to assist in project development and capacity-building for national institutions in Uganda. ECCM undertook two scoping exercises and made the following recommendations:

1. The most effective means of delivering a carbon-offset project would be through an existing national not-for-profit, non-government organization (NGO) or through setting up a new national body. This body/organization would be charged with coordination, registration and marketing of carbon credits; act as a 'clearing house' and a point of contact between farmers and potential investors in carbon credits; and provide an administrative structure to coordinate and administer carbon-offset projects in Uganda. The pilot project would be

    coordinated through an existing national conservation NGO. For this role, ECCM recommended the Environment Conservation Trust of Uganda (EcoTrust), a national conservation NGO.

2.   A land-use-based NGO would need to be involved in helping farmers with provision of technical skills for the establishment of trees and woodlots and to build their capacity in monitoring their carbon 'accounts'. The NGO would be involved in community work and forestry activities and would provide the first point of contact between the farmers and the carbon management project. For this role, CARE Uganda was deemed appropriate, based on its experience of working with farmers in the earmarked pilot districts.[3]

3.   An institution needed to be identified that would produce technical specifications, establish carbon baselines, model predicted carbon off-takes and provide monitoring standards for the pilot project. The World Agroforestry Centre, the Forest Research Institute of Uganda and the National Biomass Centre were deemed the most appropriate, based on their experience and technical expertise in these areas.

As one of the means to implement its new mandate, EcoTrust established the Trees for Global Benefits Programme as the project to pilot the new framework and lay the foundations for subsequent replication in other sites.

TREES FOR GLOBAL BENEFITS PROGRAMME

The Trees for Global Benefits Programme aims to develop and operation-alize a model for carbon trading which could be replicated in other parts of Uganda. It also aims to enhance the institutional and technical capacity of participating institutions, and to establish an institutional structure for administering Land Use, Land-Use Change and Forestry (LULUCF)[4] projects for carbon trading.

    In practice, the programme supports smallholder farmers in western Uganda to plant trees for carbon sequestration and subsequently receive payments for the associated carbon credits. It operates in three districts of Uganda (Bushenyi, Hoima and Masindi, see Figure 7.1), and the sellers of ecosystem services are individual farmers or farmer groups that own land or have user rights, often in the form of long-term leases over private land or national forest reserves. Farmers are contracted to grow trees on their land for 25 years for short rotation trees and 50 years for long rota-tion trees, and are paid according to the estimated tons of carbon dioxide sequestered.

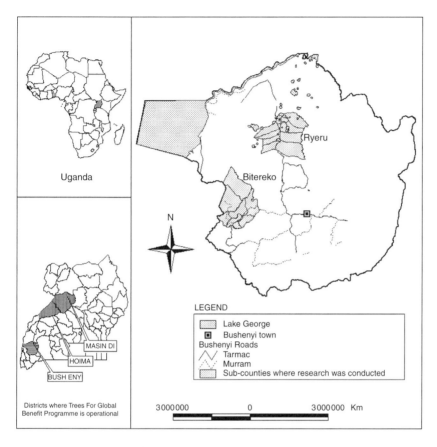

Figure 7.1   Map of the study area

Implementation is largely based on the Plan Vivo approach, which is a set of standards, processes and tools to develop and register payments for ecosystem services (PES) projects in developing countries. Plan Vivo provides a framework for managing the supply of verifiable emission reductions from rural communities in a way that is beneficial to rural livelihoods and the environment. In addition to carbon sequestration, it requires all projects to provide additional local social and environmental co-benefits to communities through the diversification of income sources, the development of sustainable land-use systems and livelihoods and planting of native species. Activities covered under the Plan Vivo system include afforestation, agroforestry, forest conservation, forest restoration and avoided deforestation.

The Plan Vivo model operates according to the following parameters:

- cash payments are made on the basis of tons of carbon sequestered per hectare by mature trees;
- each seller must establish a minimum of 400 trees;
- all trees planted must be indigenous African species or selected fruit trees;
- timber harvesting must not occur before 20 years;
- payments are staggered as follows:
  - year 0: 30 per cent of total payment, following verification that 50 per cent of the intended number of trees have been planted;
  - year 1: 20 per cent of the total payment, following verification that 100 per cent of the intended number of trees have been planted and 100 per cent have survived;
  - year 3: 20 per cent of the total payment, following verification of a minimum 85 per cent survival rate;
  - year 5: up to 20 per cent of the total payment, pro-rated according to (tree) growth performance as measured against technical specifications (anticipated growth rates for different species);
  - year 10: up to 10 per cent of the total payment, pro-rated according to growth performance as measured against technical specifications.
- third-party verification occurs at year 5.

BioClimate Research and Development (BR&D), a non-profit organization, was responsible for the development and maintenance of the Plan Vivo system and contracted the Edinburgh Centre for Carbon Management (ECCM) to assist in project development and capacity-building, particularly in the application of the Plan Vivo model. ECCM is also contracted to source buyers. In 2008, the Plan Vivo Foundation took over the management of the Plan Vivo programme from BR&D; ECCM has retained its role.

EcoTrust is charged with project coordination, farmer registration and the marketing of carbon credits from the project. It acts as a 'clearing house' and a point of contact between farmers and potential investors in carbon credits, and provides an administrative structure to coordinate and administer the carbon-offset project. EcoTrust often has direct contact with buyers, but its ability to negotiate prices is limited, as prices are generally set by global market conditions.

The programme currently involves approximately 200 carbon sellers (with a further 200 applications being processed) and the earliest plantations were established in 2003. No new farmers are accepted into the programme until buyers are secured, although supply – in the form of numbers of farmers who have formally applied – currently exceeds demand. Given the limited ability of the programme and the market to accommodate all interested sellers, a set of selection criteria is used to screen potential participants. Tenure security is a pre-condition for involvement and farmers living close to protected areas are given priority as a means to reduce harvesting pressure on that protected area, by providing alternative sources of fuel wood. Outside of these criteria, sellers are selected on a first come, first served basis.

Tenure verification does not require a title deed, which is rare for most rural households. Rather, local authorities help to verify ownership through records of prior land purchase agreements and wills. With the exception of land leased from national forest reserves (which is public land), most land is held by individual households. While contracts for carbon sequestration are signed with individual farmers, training and monitoring is carried out within farmer groups.

According to EcoTrust staff, the main transaction costs for EcoTrust are experienced at the stages of capacity-building, planning (to develop 'living plans' for how trees will be planted on each farm) and monitoring, and the up-front costs of developing technical specifications. For farmers, the main transaction costs come in the form of time spent in training, monitoring, meetings and hosting visitors interested in witnessing the pioneers of carbon forestry in Uganda. All training is hands-on and includes nursery management, planting and plantation management. ECCM assumes all transaction costs associated with securing buyers.

EcoTrust staff believe that the main constraint to finding buyers is the failure to pursue certification by either the Voluntary Carbon Standard[5] (VCS) or the Climate, Community and Biodiversity Standard[6] (CCBS) – given that these standards provide guarantees[7] to buyers. Both sets of standards were designed to ensure that carbon offsets available for purchase can be trusted and have real mitigation impacts, while promoting an additional set of co-benefits. For example, CCBS standards were designed to ensure that land management projects simultaneously minimize climate change, support sustainable development and conserve biodiversity. However, while certification is expected to open up the market both in terms of price and volume of buyers, it is currently considered too costly a process for many organizations, including EcoTrust.[8] Due to the transaction costs of qualifying many smallholder growers, EcoTrust considers certification relatively unviable for smallholder afforestation or reforestation projects.

Technical specifications for measuring carbon sequestered and for plantation establishment were developed based on the estimated amount of carbon in each species at maturity and the cultivation system to be used – whether woodlots, intercropping or boundary planting. The cultivation system influences recommendations for spacing and management, as follows:

1. Woodlots: assuming a spacing of 5 × 5m, 400 trees are required to meet the minimum 1 ha limit.
2. Intercropping: larger spacing recommended; trees are counted to derive hectare equivalent (minimum of 400 trees per seller).
3. Farm boundaries: 5m spacing recommended; trees are counted to derive hectare equivalent (minimum of 400 trees per seller).

Rather than use a per species system of measurement, EcoTrust assumes an average per-hectare value (which is in the order of US$500 per ha and a market price of US$8 per ton). The price varies from US$8 to US$14 per ton of carbon dioxide sequestered, depending on the buyer.[9] According to EcoTrust, payments amounted to approximately US$300000, as of December 2007.

In the early stages of the project, farmers received 57 per cent of total revenues from carbon sales, while EcoTrust and BR&D received 29 per cent and 14 per cent, respectively. In 2008, changes to the structure and management of the programme stimulated a shift in the allocation of revenues amongst stakeholders. The recently established Carbon Community Fund is now allocated 6 per cent of carbon sale revenues and participating farmers can access money from the Fund for project-related activities, for example, to establish a community nursery, for additional training in tree planting and management. The Plan Vivo Foundation receives 5 per cent for managing the Plan Vivo programme (which it took over from BR&D) and 10 per cent is allocated to project verification costs. Farmers now receive 51 per cent of carbon sale revenues and EcoTrust continues to receive 29 per cent. However, the funds that EcoTrust receives from carbon sales are insufficient to cover the full cost of administering the project, which therefore requires additional subsidies by EcoTrust and its other funders. Project sustainability therefore rests on ongoing financial support to EcoTrust.

Monitoring is conducted by local project coordinators, and payments made if farmers reach their targets. This regular monitoring places a strong emphasis on tree performance (survival and growth rates), although periodic meetings between EcoTrust and farmers enable emerging challenges to be identified and addressed. Several monitoring visits were also carried out by BR&D, using a combination of qualitative observations in the field and focus group discussions with farmers (BR&D and EcoTrust 2006; BR&D 2007).

According to EcoTrust, there have not been any cases of non-compliance with contracts. If farmers are not able to meet the growth requirements for the year, they forfeit payment and are given an opportunity to try again the following year. Theoretically, if a farmer were to cut down the trees after the final payment, he or she would have to pay back the money received. However, the incentive for the farmer to keep the trees standing is quite high, as they plan to sell timber in the future.

To date, there has been little third-party verification[10] or monitoring of wider social and environmental impacts, although EcoTrust does wish to pursue third-party verification in order to enhance the credibility of the project and strengthen its position in the market (BR&D 2007). The primary environmental aim of the programme – reduced pressure on adjacent protected areas – does not seem to be the subject of monitoring by the programme.

METHODOLOGY

Research was carried out in Bushenyi District, where the Trees for Global Benefits Programme has been operating longest, and focused on two of the four sub-counties covered by the programme: Bitereko and Ryeru (see Figure 7.1). Located in south-western Uganda between latitudes 0.2°–0.5° south and longitudes 25.8° and 30.7° east, Bushenyi has a total population of 731 393 people (3 per cent of the national population). The district covers a land area of 3949 km², 56 per cent of which is cultivable (arable), 20 per cent is tropical high forest, 10 per cent national parks, 9 per cent open water and 5 per cent wetlands (BDLG 2004). Most of the district is hilly, with the Rift Valley and the plateau as the other relief features. The altitude ranges between 744 and 1116 m above sea level. Agriculture may be characterized as a mixed crop–livestock system dominated by banana, the predominant cash and staple crop. Other crops include maize, sweet potato, beans, tea and coffee. Livestock are few in number and include small and large ruminants and poultry. Cattle are mainly grazed on private, fenced farms. Trees are largely grown in small woodlots on farms, most of which are less than 1 ha in size. The small size of average land-holdings (6.6 ha for participating households, 2.6 ha for non-participating households) is one of the main challenges facing the introduction of a carbon-offset programme into the district, given the competition with food crops.

Data were collected using key informant interviews, focus group discussions and household surveys. Key informant interviews were held with the District Head of the Department of Natural Resources, the District

Forestry Officer and the EcoTrust head office in Kampala, to gain an understanding of the key design features of the PES project. Focus group discussions were then held in two different parishes, with groups of carbon sellers and groups of non-participants (neighbouring farmers and others). The purpose of these focus group discussions was to identify the primary outcomes and impacts observed (both positive and negative), changing land-use patterns, support services seen as essential to enabling participation, barriers to participation, patterns in local benefits capture and primary uses of carbon payments. The purpose of these meetings was also to 'ground truth' the household surveys by incorporating variables seen as important locally (that is, types of social and environmental impacts). In addition to addressing the main questions guiding the research, household surveys assisted in quantifying qualitative observations derived from the focus group discussions.

A total of 60 household interviews were conducted, roughly half of these with participating farmers ('carbon sellers') and half with neighbouring (non-participating) farmers. Sampling of participating farmers was done randomly at sub-county level, using lists of all participating farmers. Sampling of non-participants was done by identifying neighbours of each participant household interviewed, in order to identify the diverse spin-offs to adjacent households (for example, those which are determined by proximity of residence and otherwise).

RESEARCH FINDINGS

Environmental Impacts

Environmental impacts were assessed primarily through farmer interviews. While project reports mention 'higher standards of land management' and 'improved environments' as beneficial outcomes (BR&D and EcoTrust 2006), the indicators used to assess these outcomes are unclear. Furthermore, farmers tend to express environmental impacts as livelihood impacts, both in this research and in monitoring reports by EcoTrust and BR&D (BR&D and EcoTrust 2006). It is difficult to assess whether environmental impacts expressed by farmers are based on actual observation or whether perception is influenced by the training received and discourse around the benefits of afforestation, given the long history of government-sponsored afforestation campaigns in the region. Observed impacts are nevertheless worth mentioning. Farmers stated improved crop productivity as a result of better land-management techniques (BR&D and EcoTrust 2006). Farmers also perceived microclimatic changes that they viewed as favourable, such as cooler temperatures and more regular rainfall. They also believed that air was fresher, due to the presence of forests. In some cases, trees have provided an important windbreak, reducing the negative impact of high winds on homes and banana plantations. Farmers anticipate that the cultivation of indigenous trees will create conditions similar to those of natural forest, though a number of these expected outcomes have not yet been seen. One of the anticipated positive benefits (as yet unseen) is increased water flows in streams. Other anticipated ecological impacts identified through focus group discussions are largely negative: increased incidence of mosquitoes (already observed), wild animals (not yet observed) and birds (already observed, and feared for their impact on crops).

Livelihood Impacts

Livelihood impacts on carbon sellers

It is important to begin with the positive benefits observed by participants, because in the early years of the project, when carbon payments are highest, these seem to outweigh negative impacts. Evidence of this is found in the responses to the household questionnaire and in the number of non-participants eagerly awaiting the opportunity to engage in the carbon market.

The most important benefit is seen to be cash income – within one year of planting, farmers are paid 50 per cent of the total PES payment – equivalent to US$250 for the majority of households that cultivate the minimum area or number of trees. With the average rural household income in western Uganda at 170 891 shillings (US$94[11]) in 2005–06, this is significant. Also significant is the lump-sum nature of the payments, as rural households are rarely able to save this much money to enable sizable purchases or investments. This enables households to make purchases or investments that otherwise would have been difficult. This money has been spent on plantation establishment costs (including purchase of seedlings), debt cancellation, payment of medical bills, purchase of furniture and home improvement and, in one or two cases, the purchase of land. Some carbon sellers used their income to pay school fees, an expenditure which is likely to have been made anyway, but which nevertheless enabled money to be used for other purposes. Some farmers invested in productive activities, including carbon (for example, tree seedlings, plantation expansion) and livestock purchases (small ruminants and cattle). Seventy-six per cent of carbon sellers claimed to have invested their income primarily in their plantation. Nine per cent and 29 per cent of farmers stated that the payment of school fees was their primary and secondary investment, respectively, while other farm investments were the third most important

Table 7.1   Carbon sellers reporting identified project impacts

Outcome or impact	Households affected (%)
Positive impacts	
Increased knowledge of different tree species and their uses	94
Increased skills (tree planting and management)	91
Improved social status in community	69
Improved microclimate	60
Reduced wind damage	49
Better connected to external institutions	43
Complementary income-generating activities (for example, beekeeping, passion fruit)	40
Increased shade for livestock	40
Improved access to firewood	37
Negative impacts	
Neighbours harm trees (for instance, grazing, fuel wood collection)	51
Trees outcompete agricultural crops	14
Mismanagement of carbon payments	6
Increased prevalence of disagreements among family members	56
Men and women affected differently by project	3

use of income. Members of focus groups for both participants and non-participants felt that carbon sellers have developed economically. Non-participants said they were motivated by what they had seen in terms of purchasing power.

A host of additional socioeconomic benefits to participating households was also identified (see Table 7.1). Increased knowledge of indigenous tree species and their uses came out strongly. Farmers indicated that, at the beginning of the project, they would plant indigenous trees without considering their associated uses. This changed through their involvement with the project, and they now plant more medicinal and high-value timber species, among others. Farmers' silvicultural skills have also improved. Indigenous trees (unlike frequently grown eucalypts) need a lot of care, from appropriate spacing to protection against browsing livestock, skills which farmers now proudly claim to have acquired. The survival rate of the trees planted recently was reportedly much higher than at the beginning of the project, a result of experiences acquired with previously unfamiliar or uncultivated species.

Complementary income-generating activities such as beekeeping and the cultivation of passion fruit, increased shade for livestock and the increased availability of firewood (from tree branches) are additional benefits experienced by up to 40 per cent of participating households.

A number of intangible benefits were also cited, such as improved knowledge of different tree species and their uses, new skills in nursery and plantation management, improvements in participants' social status within society, favourable microclimatic changes and enhanced 'connectedness' to external institutions. Improved social status locally was largely due to participants' economic success in entering the carbon market and created jealousies among their neighbours, as well as the perceived benefits of being known worldwide for their involvement in carbon trading. Farmers also claimed that social capital ('the spirit of togetherness') had been enhanced among tree farmers and there was greater 'self control' in sustaining rather than cutting down trees for short-term benefit.

Additional benefits mentioned in focus groups (but not captured by household surveys) included income and nutritional benefits from the cultivation of fruit trees, the increased productivity of banana from transferring biomass (slashed undergrowth) to banana plantations, improved productivity of pasture when trees are integrated into the system and improved awareness of climate change. The tendency to 'encroach on other people's gardens' was also said to have reduced due to the increased availability of firewood. When asked whether the project had any influence on the way in which people related to the environment, farmers stated that, before the programme, people did not understand the importance of afforestation or the value of indigenous trees; 'people wanted to see flat land, just for farming.' Farmers claimed that radio campaigns, personal experience and the project itself have enhanced farmer awareness of how land degradation affects crop yields and of the benefits of farm forestry.

There are, however, significant opportunity costs and risks for carbon sellers that may arise from the loss of alternative economic activities. During focus group discussions, farmers complained that the carbon project had reduced crop yields and income for some households, as trees had been planted on land formerly used for cash crops, pasture or existing woodlots. Households that have attempted to intercrop with bananas or annuals claim that, within a few years, the trees outcompete the crops and intercropping is no longer viable. For a few households, poor tree growth, perceived to be due to disease or excess sunshine, have resulted in failed investments. Forty-four per cent of households have planted carbon trees on what was formerly cropland, 35 per cent on pasture land, 29 per cent on degraded land and 15 per cent in already existing woodlots. A number of farmers complained about the effect this has had on the productivity of cropland, grazing land and eucalypt woodlots. This supports the

Table 7.2   Average reduction in agricultural production resulting from the establishment of carbon plantations, and corresponding market value

Activities displaced by carbon plantations	Average amount	Average value ($US)
Reduction in crop harvested from area:		
Banana (bunches/year)	63	56.63
Beans (kg/year)	19	7.97
Cassava (kg/year)	136	24.45
Coffee (kg/year)	56	40.23
Groundnut (kg/year)	14	12.44
Millet (kg/year)	21	11.59
Pineapple (fruits/year)	15	2.65
Sorghum (kg/year)	7	2.12
Sweet potato (kg/year)	64	11.56
Reduction in number of livestock grazing in area:		
Sheep/goats (head at any given time)	2	n/a
Cattle (head at any given time)	1	n/a
Reduction in forest products derived from area due to competition or felling:		
Fuelwood (bundles/year)	18	10.58
Poles (number in average year)	23	24.37
Number of trees felled for establishment of carbon plantation	15	n/a

observation that, despite the voluntary nature of involvement, opportunity costs associated with displaced land uses and transaction costs may be high (see, for example, Pagiola et al. 2005; Corbera et al. 2007).

Table 7.2 summarizes these costs through household averages. While households differ greatly in the activities they choose to displace with carbon trees, average values are helpful in illuminating the net economic effects of the programme.

Out of pocket expenditures related to displaced land uses were calculated by asking participants if the displaced land uses required them to purchase anything they used to produce themselves. If they said yes, each product was recorded and participants were asked to quantify how much they purchased before and after the shift to carbon forestry. The difference was converted into an expenditure based on the per unit cost. While three out of seven households in one focus group indicated that they now bought crops they used to cultivate, only 24 per cent of survey respondents

Table 7.3   Expenditures related to displaced land uses

Product	Expenditures ($US/year)	Standard Deviation
Crops		
Beans	1.41	1.41
Cassava	0.73	0.71
Groundnut	2.82	2.82
Millet	3.18	2.34
Livestock		
Milk	7.62	5.31
Total	15.76	–

indicated that displaced land uses resulted in additional expenditures on food. The number of households purchasing displaced commodities varied by product, but with only 3–6 per cent of surveyed households purchasing any given item, it would seem that the vast majority of households forego consumption rather than purchase displaced products. The average cost implications of additional food expenditures are described in Table 7.3. Displaced activities also lead to non-food expenditures, such as the purchase of fuelwood or dung otherwise available on-farm.

It is difficult, with the available data, to estimate the net economic effects on households of adopting carbon forestry practices, though indications are that the costs of displaced production and additional expenditure on food items may outweigh carbon income, particularly beyond the initially higher payments. Some areas of uncertainty in calculating financial impacts include: the cost implications of displaced livestock activities and future income from timber sales in carbon plantations. The highest costs to households would seem to be in years 4 to 20, when income from crops has declined but there are limited or no carbon-offset payments to compensate farmers. The primary financial benefit is thus the higher up-front payments which enable investments that would otherwise not be possible. For example, investments in (indigenous) woodlots that will ultimately yield higher returns than the displaced woodlots (largely planted with eucalypts), or investments in crop and livestock production.

The question of costs needs to be considered not only in terms of loss of other productive activities but also in terms of the need to commit that piece of land to carbon for a minimum of 20 years, which minimizes a household's ability to adapt to changing circumstances. Up-front payments have proven highly effective in luring farmers into the carbon market, but may prove to be a source of dissatisfaction in later years, particularly in the last 10 years of the contract when payments for carbon are no longer received

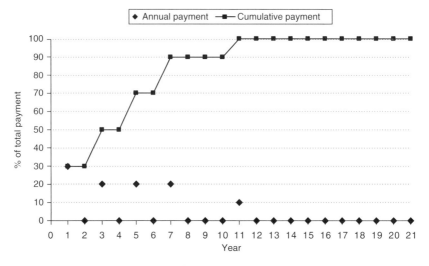

Figure 7.2   Payment schedule for EcoTrust carbon forestry project

(see Figure 7.2). While some authors (Wunder 2008) claim that convincing real world examples of 'environmental service traps' with systematic welfare losses seem to be missing, this project could well encounter a great deal of farmer discontent in its later years. EcoTrust has tried to minimize the involvement of farmers who cannot afford to do so by asking farmers, up front, if they can afford to plant trees without compromising food availability. However, this approach is unlikely to minimize the seductive lure of high up-front payments. As one farmer stated: 'where there is money, you may be forced to say there is enough [land] when there is not.'

   While it is difficult to predict the long-term consequences of the EcoTrust scheme, some authors have argued that poor sellers could become trapped in long-term land-use deals with negative livelihood outcomes (Landell-Mills and Porras 2002). At the same time, while a more even distribution of payments over the life of the plantation is possible, it is important to retain some front-loading of payments to help poorer households meet the investment costs (Wunder et al. 2008). One possibility for minimizing the opportunity costs in later years while supporting farmers with costly up-front investments would be to front load only that amount of payment that is necessary to enable participation (for example, in the form of subsidies to finance start-up costs), thus enabling less wealthy households to participate, while taking the balance and spreading them evenly over the 20-year contract period. This would have two benefits: a more honest 'lure' up front (making it easier to say 'no') and a better spread of payments later on (making it easier to stay in the market).

A second strategy applied by EcoTrust to minimize the costs borne by the household is to ensure that no more than 25 per cent of a household's total landholding is allocated to carbon forestry. Since the project requires at least 1 ha be allocated to carbon forestry, EcoTrust recommends that only those households with a minimum of 4 ha participate. Data from household surveys indicate that participating households have an average of 6.6 ha of land, compared with 2.6 ha for non-participating households. Contrary to project advice, 26 per cent of the participating households surveyed have landholdings of 3 ha or less. Households with smaller farm sizes tend to integrate trees into existing land uses (cropland, grazing land or existing woodlots) rather than replace them. Some households expressed concern that as trees mature they have begun to outcompete cropland with implications for food availability.

Farmers claim that there are a number of risks associated with participating in the carbon market, such as delayed payments from the buyers or intermediaries. A few farmers noted that the first payment did not arrive in time to enable the purchase of seedlings for meeting the requirements to receive the second payment (that is, 100 per cent of intended area planted). The state of the global economy also poses risks, for instance, carbon is priced in US dollars and the value of the dollar declined by about 17 per cent between project inception and the time of fieldwork (in September 2008). Farmers are paid the Ugandan shilling equivalent of the dollar at the time of payment, and currency fluctuations create some mistrust towards project managers. Other farmers feel that their knowledge is insufficient to establish and manage plantations of indigenous tree species effectively (for example, raising seedlings, transplanting, thinning, pruning).

Neighbouring (non-participating) households and children are also said to pose risks to newly established plantations, damaging trees by gathering firewood, allowing their livestock to graze freely or tying livestock to young trees. Perhaps the most significant risks are perceived to be those posed by the government and by carbon buyers. Farmers fear that government policies prohibiting the felling of indigenous trees on public land – intended to ensure the integrity of protected areas – will affect them when they try to sell their timber. Yet the *National Forestry and Tree Planting Act, 2003* provides for the establishment of private forest plantations, specifies that all forest produce derived from such private plantations belongs to the owner and that the owner may in turn utilize this produce as he or she wishes.[12] There is also a widespread fear that their land will be 'taken away by whites' who are buying their carbon. When exploring the reasons for this fear, focus group participants simply stated that people 'do not fully understand what selling carbon means.' This comment does not seem to apply to the terms of the agreements, but rather to the farmers' inability to comprehend why someone would want to pay them for keeping trees standing that they themselves will sell and from which they will profit. No existing frame of reference helps them to interpret benefits which are otherwise not visible to them.

Livelihood impacts for neighbouring farmers and other non-sellers

While not immediately evident, a decision by one household to invest in carbon plantations has a host of spillover effects on neighbouring households and the community at large. Some of these effects are positive, such as casual employment in plantations (for example, slashing), increased local demand for agricultural products due to increased income among carbon sellers (and perhaps also due to their inability to cultivate as much as before), increased availability of flowering trees for beekeepers and improved availability of botanical medicines.

A number of negative impacts were also identified, associated with shifts in customary patterns of resource access and tenure. People once had access to land which sat idle or was fallow to gather fuelwood, to graze livestock or borrow for cultivation purposes. As stated by one farmer, 'here the custom goes that if anyone has free land, everyone can use it.' Seeing that income can be generated from such land has motivated the children of growers to get involved in woodlot establishment on their parents' land, and motivated participating farmers to expand their area planted to carbon forestry. As a result, customary access to land by neighbouring farmers (non-participants) has become increasingly restricted. Some households have had to hire land to grow crops, where once they would have grown them on land accessed freely, as the 'underutilized' land once lent out to those in need has now been diverted to carbon forestry. The costs associated with land rental depend on the crop to be grown.[13] The restricted availability of unutilized land has been particularly hard on those families with insufficient cultivable land of their own, who must now buy their food.

Almost one-third of non-participating households interviewed claimed to have suffered reduced access to agricultural land as a result of the expansion of carbon forestry. This supports the observation by Grieg-Gran et al. (2005) that since PES schemes are designed for land-owning and 'land controlling' households, they are inherently less suitable to assist the poorest of the poor. It also supports the caution expressed by Kerr (2002) that the livelihoods of the landless poor, who often depend on customary use rights to land to meet essential food, fodder and fuel needs, may be harmed if PES conditions limit their access to previously 'unproductive' land.

Before the arrival of the Trees for Global Benefits Programme, goats were said to graze freely but are now prohibited, due to their potentially

Table 7.4   Number of non-participating households identifying negative impacts of the Trees for Global Benefits programme

Type of outcome or impact	% Households affected (n=30)
Negative impacts	
Lost access to grazing area	36
Restricted access to agricultural land (on loan)	32
Increased prevalence of mosquitoes	28
Shading/root competition with neighbouring crops	28
Lost access to fuel wood	16
Restricted access to sand for construction	8
Positive impacts	
Employment in tree plantations	12
Those growing trees are buying more of our agricultural produce	8
Increased honey production due to increase in number of trees	4

detrimental impact on young plantations. This has led to a number of disputes between households, with conflicts forwarded on to the Local Council or to EcoTrust for resolution. Gathering of fuelwood and construction materials (poles, sand) on neighbours' plots has also been curtailed for some households.

In addition to the shift in customary practices, there are other trans-boundary effects which impact on neighbouring households (see Table 7.4). Some farmers have observed trees competing with their cropland for nutrients, moisture ('drying the soil') and sunlight. In some areas, the prevalence of mosquitoes was said to have increased due to microclimatic changes. Certain tree species planted in carbon forestry woodlots such as Munyamazi (*Rauvolfia vomitoria*) are also said to emit an offensive smell and reduce the quality of honey.[14] This problem is restricted to the few households engaged in apiculture, but nevertheless deserves attention, as EcoTrust is promoting apiculture as a complementary economic activity to carbon forestry. Neighbouring farmers are also concerned about damage to their cropland at the time of tree-felling, an issue which was raised by some households but not included in the household surveys.

Catalysing a dialogue among beneficiaries (carbon sellers) and neighbours or other households negatively affected by the scheme can be a way of exploring options for enhancing the wins and minimizing the losses to neighbouring households. Previous experience suggests that using such a 'multi-stakeholder' approach to identifying socially optimal outcomes, combined with local by-laws to support the implementation of

Table 7.5   Major barriers to participation faced by non-participating households

Barrier	Level of Importance (on a scale of 1-10)*
Small landholdings (farm size)	6.7
Competition with more important land uses	6.3
Limited capital to buy seedlings and prepare land	5.4
Lack of information about the project	5.2
Hesitance to commit land to forest or delay harvest for 20 years	5.1
Fear that land will be taken away by the buyers	4.6
Fear that they will be unable to harvest indigenous trees	4.5
Fear they will not find a buyer	3.6
Fear that forest will attract mosquitoes or wild animals	3.7
Failure to reach consensus within the household	2.7

Notes: * Farmers rated the extent to which each factor affected their ability to participate where 1 meant a factor was least likely to be a barrier and 10 meant it was most likely to be a barrier.

resolutions, can go a long way in fostering cooperation for mutual benefit arrangements at landscape scale (German et al. 2008).

Distribution of benefits

The most significant factor influencing the flow of benefits from carbon forestry was said to be wealth: the availability of land and capital were said to be a strong determinant of a household's ability to participate in, and invest in, carbon forestry. The average landholding of participating farmers was found to be 6.6 ha, compared with just 2.6 ha for non-participants (with 47 per cent of participating families having at least 6 ha of farmland). This supports the observation by Nowak (1987) that larger holdings may be more able to adopt PES-promoted land uses than smallholders with a strong subsistence orientation, given the trade-offs associated with meeting food security needs. One farmer said that the project had improved the livelihoods of both the upper and middle classes, as those with larger plots simply planted more trees and thus benefited more relative to those with smaller plots (and thus fewer trees). Small landholdings and the competition of trees with more important land uses (for instance, those associated with food security) therefore stand out as the major barriers to participation (see Table 7.5). While the project's recommendation of a maximum of 25 per cent of a household's landholdings allocated to carbon farming could be seen as a reason why households with limited land area tend not to participate, this requirement seems to be necessary for minimizing the burden on household food security. Poorer households who have joined the project were said to have made difficult choices on the activities to be displaced, such as food crops (millet, sweet potatoes), which they then have to buy.

Capital is also required in order to participate in, and benefit from, the programme. A minimum initial investment of approximately US$96 (for 400 seedlings, US$0.18–0.30 per seedling) is necessary, as are additional costs associated with maintaining plantations. Another factor in the pattern of adoption was limited information flow, with those farmers able to attend meetings often the ones to plant trees, because of the technical information acquired.

Other barriers to participation included the delay in benefits associated with cultivation of indigenous trees and the delay in receiving payments which discourages others from joining. Further disincentives included the distance of the source of seedlings from most of the people who want to join, insufficient household labour and the fear of creating misunderstandings with neighbours whose land is adjacent to their plots.

Another important factor in local attitudes toward indigenous trees is the history of government-supported afforestation efforts. Governments have historically promoted cultivation of exotic trees for timber and dissuaded farmers from cultivating indigenous trees given the difficulty of determining whether the timber is derived on-farm or from protected areas. Believing the government must be a role model for others, farmers expressed frustration at the disconnect between EcoTrust's emphasis on indigenous trees and the government's continued emphasis on exotics. As stated by one farmer, 'Why are you not targeting government first? Government has not yet discovered the importance of indigenous trees. They are planting different species from what EcoTrust is promoting, and they have large forest reserves.' This expresses both a desire to see government internalizing the unique contribution of indigenous species, as well as a frustration that those with the smallest land area are being asked to integrate slower growing species into their farms.

Attempts to explore the gender implications of the project yielded several observations worth mentioning. Among participating households, men and women both claimed that carbon forestry has had no impact on the gendered division of labour at the household level. Attention to this needs to be maintained as the project expands, however, the tendency to plant trees in areas formerly used for crops (the latter being largely managed by women) would tend to reduce women's labour burden as tree growing is less labour-intensive than agricultural crops.

Men were said to dominate decisions over whether to join the programme, and over the use of income from carbon. As stated by one (male) farmer, 'culturally, the man owns the land and anything a woman does on the land must be based on prior consultation with the man.' More research is needed on the potential consequences for women and children of a shift in the main sources of income. Opinions were mixed on whether the project has contributed to any disagreements at the household level. In Bitereko sub-county, farmers stressed consensus-based decision making at the household level and the 'togetherness' induced by a women's group bringing such an important income-generating activity into the household. One woman in Rutookye claimed that many women wanted to join the project, but that their husbands would not let them. The reason given was that men feared that 'the woman will get rich quick and go ahead of him', while male participants suggested it was due to competing land uses (men preferring to raise cattle) and women's 'lack of knowledge on tree planting.' Age also seems to be a factor in the interest expressed in tree farming, with a number of youths initiating their own plantations and one elder expressing his dissatisfaction with the delayed returns from timber harvesting, given his age and the time taken to benefit from timber harvesting. The tendency for land to be inherited through the male line and the symbolism of tree planting as an act which stakes a claim to land are also likely to influence cultural attitudes toward farm forestry as a male activity.

To equalize patterns of benefit capture, it is important to consider ways to minimize the role of poverty (in particular, household land and capital endowments) in determining the extent to which households can benefit from emerging carbon markets. One possibility may be to explore local institutional arrangements for 'bundling' households to produce the required minimum units of the service (that is, one hectare or 400 trees), combining local institutional arrangements with bylaws for governing individual behaviour (for example, rules on distribution of rights and responsibilities, as well as sanctions for non-compliance). Another possibility may be for land-limited households to gain access to under-utilized or under-performing public or private land (for instance, national forest reserves, absentee landholders), with emphasis on mutual benefit arrangements.[15] Payment schemes or credit systems to minimize economic barriers related to start-up costs may also be a necessary complement to the issue of land shortage for enabling poorer households to capture benefits.

Support Services

Participants and non-participants were asked to identify the most critical support services required to enable a household to participate in the carbon market. Farmers mentioned both the services received and those

Table 7.6   Perceived importance of different support services for participation in the voluntary carbon market for participating and non-participating farmers

Support services	Average ranking*
Information about how to calculate payments	8.5
Awareness about how the carbon market works	8.2
Credit for plantation establishment	7.2
Training in species selection	7.0
Training in complementary economic activities	6.9
Awareness creation about relevant national laws and how they affect tree harvesting	6.9
Training in how to plant and manage plantations	6.7
Training in nursery management	6.1
Farmer exchange visits	5.9
Support to get title to their land	5.8
Negotiating access to protected areas to acquire planting material	4.8

Notes: * Rated the importance of services on a scale of 1 (least important) to 10 (most important).

desired, and these services were then ranked by participating and non-participating households. Table 7.6 shows that the most crucial support services were information about how to calculate payments and general awareness about the carbon market: who the buyers are, and why they are interested in carbon.

The main issue raised by farmers with regard to calculating payments was their inability to independently verify whether their payments were correctly calculated and paid. Farmers expressed concern that they did not understand the contents of the individual carbon sale agreements signed with EcoTrust. These agreements are prepared in English, even though most of the farmers do not speak English. Though EcoTrust took them through the agreements during recruitment and training, few of them really understand the content of these contracts. The translation of the agreement into the local language is therefore necessary, so that farmers can fully understand all the terms therein, especially how much they are paid over the project's lifetime. It is equally important for them to be able to access current exchange rate information and to understand, ahead of time, the risks associated with exchange rate fluctuations. However, all identified services are perceived to be important. Credit or up-front payments were also seen as a fundamental service to facilitate purchase of seedlings and/or nursery operation and land preparation.

CONCLUSIONS

Stakeholder perceptions about the carbon forestry project being piloted by EcoTrust in Bushenyi and other districts of Uganda were generally very positive. Most farmers – both participants and non-participants – see this as a unique opportunity to generate significant revenue relatively quickly, enabling them to realize goals that would otherwise have been difficult to achieve. A key indication of this success is that the list of farmers seeking to become involved is large and growing. However, there are also a host of barriers constraining the poorest households from benefiting, and evidence to suggest that risks to livelihood exist for households that 'can't resist' the initial payment, but for whom the opportunity costs of allocating significant portions of their landholdings to carbon trees are likely to increase over time as carbon payments decline and other activities are increasingly out-competed by trees. A host of complex interactions and spin-offs, both positive and negative, were also shown to characterize local landscapes where this project is operating, such as a reduction in customary 'safety nets' (for instance, land access by poor households), microclimatic effects (decreased wind, increase in mosquitoes) and transboundary effects (for example, competition with adjacent cropland). To further consolidate the positive gains from this project, innovations to address these shortcomings should be developed and tested.

NOTES

1. For more details, see the National Forestry Authority website, www.nfa.org.
2. While the report attributed this to the high dependency of the rural poor on forest products, the literature has also shown this to be due to limited institutional capacity to enforce rules of exclusion in public lands.
3. The pilot districts included Bushenyi and Kasese in western Uganda.
4. The United Nations Framework Convention on Climate Change defines LULUCF as 'a greenhouse gas inventory sector that covers emissions and removals of greenhouse gases resulting from direct human-induced land use, land-use change and forestry activities' (http://unfccc.int/essential_background/glossary/items/3666.php#L).
5. The founding partners of the VCS are The Climate Group, the International Emissions Trading Association and the World Business Council for Sustainable Development.
6. The CCBS is managed by the Climate, Community and Biodiversity Alliance, a group that acts as a regulating body and whose members include conservation organizations and corporations and advising institutions.
7. For example, that the carbon sequestered through the project is additional (that benefits would not have been achieved in the absence of the project), measurable, permanent and unique (not used more than once to offset emissions); that no leakage has occurred (emissions shifting elsewhere as a result of the project); that a set of social and environmental co-benefits has been achieved; and that these impacts are independently verified.
8. Project validation and subsequent verifications with the CCBS are estimated to

range between US$5000–40000, while combined Carbon Fix Standard/CCBS cer-
tification is estimated to cost around US$28000. VCS validation and verification is
estimated to range between US$15000–30000 for each third-party audit. Plan Vivo
validation, on the other hand, costs between US$5000–12500 and verification between
US$15000–30000 (Merger 2008).
9. Most large buyers (for example, Tetra Pak, Future Forests) pay US$8 per ton of CO_2,
while individual buyers, especially those seeking to offset carbon emissions from their
travel, are often willing to pay more.
10. Third-party verification was carried out by the Rainforest Alliance in October 2008, but
the report is not yet finalized or available to the public (Heyward, pers. comm.).
11. http://www.oanda.com/convert/fxhistory.
12. The Act makes no distinction between indigenous and exotic species in this regard.
13. According to one farmer in Bitereko, she pays 10000 Uganda shillings to hire 0.25ha
for potato, but pays 20000 Uganda shillings for growing cassava on the same size of
land and 30000 Uganda shillings for millet.
14. The individual raising this concern participated in one of the focus group discussions,
but not in the household survey.
15. For example, national forest reserves are often under-performing relative to both social
and ecological criteria, given limited capacity to enforce rights to exclusion.

REFERENCES

BioClimate Research and Development (BR&D) (2007), Trees for global benefit,
Plan Vivo Field Trip Report, 24 November to 3 December 2007, Edinburgh:
BR&D.
BioClimate Research and Development and EcoTrust (2006), Trees for global
benefit: a Plan Vivo Project, Bushenyi Field Trip Report, Uganda, 17 to 20 July
2006, Edinburgh and Kampala: BR&D and EcoTrust.
Bushenyi District Local Government (BDLG) (2004), District State of Environment
Report, Bushenyi, Uganda: BDLC.
Corbera, E., K. Brown and W.N. Adger (2007), The equity and legitimacy of
markets for ecosystem services, *Development and Change* 38 (4), 587–613.
German, L., W. Mazengia, W. Tirwomwe, S. Ayele, J. Tanui, S. Nyangas, L.
Begashaw, H. Taye, Z. Admassu, M. Tsegaye, F. Alinyo, A. Mekonnen, K.
Aberra, A. Chemangeni, W. Cheptegei, T. Tolera, Z. Jotte and K. Bedane
(2008), Enabling equitable collective action and policy change for poverty
reduction and improved natural resource management in the Eastern African
Highlands. CAPRi Working Paper 86, Washington DC: IFPRI.
Grieg-Gran, M.A., I. Porras and S. Wunder (2005), How can market mechanisms
for forest environmental services help the poor? Preliminary lessons from Latin
America, *World Development* 33 (9), 1511–27.
Kerr, J. (2002), Watershed development, environmental services, and poverty alle-
viation in India, *World Development* 30 (8), 1387–1400.
Landell-Mills, N. and I. Porras (2002), *Silver Bullet or Fools' Gold? A Global
Review of Markets for Forest Environmental Services and their Impact on the
Poor*, London: International Institute for Environment and Development.
Merger, E. (2008), Forestry carbon standards 2008: A comparison of the leading
standards in the voluntary carbon market and the state of climate forestation
projects, available at http://www.carbonpositive.net/ (accessed 15 February
2009).

Ministry of Water and Environment (MWE) (2007), *Environment and Natural Resources Sector Investment Plan*, Kampala: MWE.

Nowak, P.J. (1987), The adoption of agricultural conservation technologies: economic and diffusion explanations, *Rural Sociology* **52** (2), 208–20.

Pagiola, S., A. Arcenas and G. Platais (2004), Can payments for environmental services help reduce poverty? An exploration of the issues and the evidence to date from Latin America, *World Development* **33** (2), 237–53.

Welch Divine, M. (2004), Three communities, two corporations, one forest: forest resource use and conflict, Mabira Forest, Uganda, Agroforestry in Landscape Mosaics Working Paper Series, World Agroforestry Centre, Yale University Tropical Resources Institute and the University of Georgia.

Wunder, S. (2008), Payments for environmental services and the poor: concepts and preliminary evidence, *Environment and Development Economics* **13**, 279–97.

Wunder, S., S. Engel and S. Pagiola (2008), Taking stock: A comparative analysis of payments for environmental services programmes in developed and developing countries, *Journal of Ecological Economics* **65**, 834–52.

8. Livelihood impacts of payments for forest carbon services: field evidence from Mozambique

Rohit Jindal

INTRODUCTION

This study reviews the livelihood impacts of the Nhambita Community Carbon Project located in Sofala Province, Mozambique. The project pays local smallholders for taking up carbon sequestration activities on their farms and for conserving miombo woodlands in the area. It follows the concept of payments for environmental services (PES), whereby land stewards receive payments for securing valuable environmental services through their conservation activities (Wunder 2005).

PES literature often highlights a potential compatibility between environmental conservation and poverty reduction, especially when poor households are contracted to receive payments in return for their conservation efforts (Pagiola et al. 2005). Such benefits are particularly significant for Mozambique, which is one of the poorest countries in the world. In 2006, its gross domestic product was only US$7.6 billion, with over 74 per cent of the population living on less than US$2 per day (World Bank 2008). Reducing this widespread poverty by investing in economic development has been a key challenge for the national government (Heltberg et al. 2003). Agriculture and forestry are two key sectors that need investment, as they employ more than 80 per cent of the country's work force and provide livelihoods for a vast majority of the rural poor (FAO 2003). Forest-based carbon mitigation projects such as the Nhambita project can potentially achieve these twin objectives – they can bring investment for improved forestry and land-use practices, as well as increase rural incomes through the provision of direct payments to local farmers (Jindal et al. 2008). However, to what extent do PES projects, particularly carbon mitigation projects, actually promote rural livelihoods in poor countries such as Mozambique? The evidence so far has been mixed, with some studies pointing to positive outcomes (for example, Miranda et al. 2003; Tipper 2002) while others highlight adverse impacts (Eraker 2000).

A major shortcoming of such studies, however, is their primary focus on anecdotal evidence. There is thus a need to conduct a non-partisan review that compares the impact on participants with that on non-participating households (Pagiola et al. 2005). Although some detailed impact studies are now available for large national level PES programmes (for example, Uchida et al. 2007), empirical evidence on the welfare effects of small-scale carbon projects remains sketchy. The present case study aims to fill this gap through a detailed investigation of the Nhambita Community Carbon Project. It answers four key questions: (1) to what extent do the local poor participate in the project, (2) to what extent do project activities generate carbon offsets, (3) what are the impact(s) on participating households, and (4) do spill-over effects impact upon non-participating households? The answers to these questions will not only provide useful information to national policy makers in Mozambique, but also contribute towards the ongoing debate over the value of forestry carbon projects for smallholders.

THE NHAMBITA COMMUNITY CARBON PROJECT

The Nhambita Community Carbon Project is located in the buffer zone[1] of the Gorongosa National Park (GNP) in Mozambique. The aim of the project is to develop forest-based land use practices that promote sustainable rural livelihoods while generating verifiable carbon emission reductions (University of Edinburgh 2002). The project began in 2003 in Nhambita village within Chicale *régulado*,[2] and since then has been extended to several neighbouring villages over an area of nearly 20 000 ha. The project is implemented by a consortium comprising the University of Edinburgh (which provides overall coordination and support for research activities), the Edinburgh Centre for Carbon Management (which prepares technical specifications for agroforestry practices) and Envirotrade Ltd (which is responsible for field-based implementation and sale of carbon credits generated by the project). During its pilot phase from 2003–08, the project was funded by the European Union (EU). Since then, the project has raised revenue through the sale of carbon offsets[3] to international buyers such as the MAN group and the Carbon Neutral Company.

There are two broad activities that produce these carbon offsets: carbon sequestration through agroforestry and reduced emissions from deforestation and forest degradation (REDD) of miombo woodlands. The first is undertaken by individual farmers on their farmlands, while the second

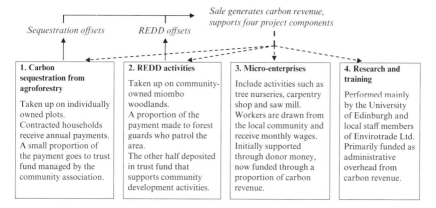

Figure 8.1   Major components of the Nhambita Community Carbon Project

is undertaken by the entire community in surrounding woodlands. The other important components of the project are the promotion of micro-enterprises, and research, extension and capacity building (see Figure 8.1).

Carbon Sequestration through Agroforestry

One of the prominent activities of the project is carbon sequestration through adoption of agroforestry by farmers on their *machambas* (farm-lands). All households within Chicale *régulado* are eligible to participate in these carbon sequestration activities. Farmers are selected on a first-come first-served basis, and are offered a 'menu' of agroforestry practices from which they choose those that suit them best. In return, the project provides them with payments for sequestering carbon. The choices include horticul-ture (in the form of cashew or mango orchards), setting up woodlots (with tree species such as *Albizia lebbeck* and *Khaya nyasica*), intercropping with *Faidherbia albida*, planting native hardwoods such as panga panga (*Millettia stuhlmannii*) around the boundary of the *machambas*, and plant-ing fruit trees such as tamarind (*Tamarindus indica*) within the homestead (see Table 8.1 for details). Prior to taking up agroforestry, most household farms are mainly used either for growing seasonal crops such as maize or sorghum, or left fallow to regain productivity. With the introduction of agroforestry practices, a certain proportion of these farms is covered by permanent trees resulting in net sequestration of atmospheric carbon for the contract period.

In order to ensure the permanence of carbon stocks, the sequestration

Table 8.1   Carbon sequestration rates under the Nhambita Community Carbon Project

	(1) Carbon sequestered (tC)	(2) Baseline carbon stock (tC)	(3) Buffer carbon stock (tC)	(4) Net carbon offsets (tC) (1) – (2) – (3)	(5) Net carbon offsets (tCO$_2$)*
Boundary planting	3.23/100 m	0	0.48/100 m	2.75/100 m	10.08/100 m
Inter-planting (*Gliricidia*)	10.00/ha	0	1.50/ha	8.50/ha	31.16/ha
Cashew orchards	40.14/ha	2.80/ha	5.60/ha	31.74/ha	116.38/ha
Mango orchards	34.00/ha	2.80/ha	4.68/ha	26.52/ha	97.24/ha
Homestead planting	42.05/ha	0	6.30/ha	35.75/ha	131.08/ha
Woodlots	61.30/ha	11.30/ha	7.50/ha	42.50/ha	155.83/ha
Inter-planting (*Faidherbia*)	58.20/ha	0	8.73/ha	49.47/ha	181.39/ha

Notes: * 1 tC = 3.67 tCO$_2$.
Estimates are based on projected tree growth under standard climatic and soil conditions, and assume that all farmers will follow a standard set of silvicultural practices. These sequestration rates may be affected by natural disasters such as prolonged drought or fire outbreaks.

Source: Tipper 2008.

rates in Table 8.1 have been estimated on the basis of maintaining tree cover for a period of 100 years. However, farmers are allowed to selectively harvest mature trees and replace them with new trees. Each agroforestry system is designated as a separate contract and usually covers 0.25–1.50 ha of *machamba* land. Once a household signs a contract, it receives free tree seedlings and training on how to plant and manage trees. For mortality rates of less than 15 per cent during the first 2 years, it receives free seedlings as replacements. Where mortality rates are higher, the household is expected to pay for additional seedlings. A household can enrol for multiple contracts, either by following the same agroforestry activity on different plots or by taking up different agroforestry activities on the same plot (for example, combining boundary planting with fruit orchards).

The Nhambita project monetizes the total number of carbon offsets generated by any new agroforestry system (calculated at an average price of US\$4.50 per tCO_2) and pays annual cash payments to contracted households.[4] It is important to note that while carbon offsets are generated over 100 years, farmers are paid the entire value of these offsets during the first 7 years of the contract. Thereafter, benefits from newly planted trees – increased land productivity, selective harvest of timber trees, availability of high value fruits and other non-timber products – are expected to provide sufficient economic incentives to the household for it to protect these trees for the next 93 years.[5] For instance, if a household takes up intercropping with *F. albida*, it receives a total of Mozambican Meticais (MZM) 20 187/ha (or US\$833) as carbon payments over the first seven years.[6] Mango plantations yield a lesser payment of MZM 10 482/ha (or \$US433) because the inter-tree distance is higher, resulting in less carbon sequestration per hectare.

An important requirement for receiving carbon payment is that contracted households do not burn their *machambas*. Burning of *machambas* to clear weeds or scare away wild animals is prevalent in the area. However, this may result in net loss of carbon to the atmosphere and therefore contracted households are expected to refrain from burning. The project team enforces this rule quite strictly. In 2007, for example, carbon payments to several farmers were withheld because they had burnt their *machambas* in the previous year and payments were only released once the health of the tress had been ascertained.

There is a high demand for these carbon-related agroforestry contracts in the area. By 2008, 852 households comprising more than 80 per cent of all households in the Chicale *régulado* had enrolled for a total of 1234 contracts. An important question regarding the sustainability of the project is whether these contracted households will actually protect the newly established trees once the cash payments have been made, after 7 years. Contracts for 100 years imply that future generations will be required to follow certain rules, with which they may not agree, particularly if any associated benefits have already been paid out. Although the agroforestry systems are expected to provide valuable benefits such as timber and non-timber products, these benefits alone may not compensate for the loss of carbon payments after the end of the first 7 years. On the other hand, reducing the contract length would reduce the number of carbon offsets and hence the payment made to a household. We return to this issue in more detail during our discussion on project impacts.

Reduced Emissions from Deforestation and Degradation

In addition to carbon sequestration, the local community receives carbon payments for REDD activities in miombo woodlands around the GNP.

Miombo woodlands are open canopy dry deciduous forests found across central and southern Africa. They are the most common forest formation within and around the GNP. These forests are dominated by trees such as *Brachystegia, Julbernardia* and *Pterocarpus* (Williams et al. 2008). In recent years, the deforestation rate of miombo woodlands around the GNP has increased due to clearance of forest for agriculture, tree-felling for charcoal production (especially along the main road), uncontrolled burning of forest and logging for timber (although most of the latter occurred in the 1980s and is now under control).

The Nhambita project motivates the local community to protect these miombo woodlands from any kind of harvesting or fire outbreak. Two kinds of activities are taken up to reduce the rates of deforestation and degradation in the area: a total ban on tree-felling and the formation of fire patrols that guard against recurrent fire outbreaks. REDD activities began in 2006 on a specially demarcated block of 5000 ha and since then have been expanded to a total of 11 071 ha. Carbon offsets are calculated on the basis of reduction in biomass loss from the average deforestation rate estimated for the entire area (Tipper 2008). Over a 10-year period, REDD activities in the area will generate 73.3 tCO_2/ha. These offsets are combined with carbon sequestration offsets from agroforestry and are sold as one lot to international buyers identified by Envirotrade Ltd. Grace (2008) estimated that, by the end of 2007, the Nhambita project had paid a total of US$223 750 to the local community in the form of carbon sequestration and REDD payments.

The actual flow of money into the community varies by activity. In the case of carbon sequestration, most of the money is paid directly to individual contract holders, while a small proportion goes into a community trust fund, managed by a democratically elected community association. The REDD payments are divided in two: one half is deposited into the community trust fund and the other half is paid as wages to the forest guards who patrol the woodlands and protect them from fire outbreaks. The community association consists of 24 members (including three women) who represent different villages in Chicale *régulado*. The association was first formed in 2001 and the second elections were held in 2006. The association meets weekly to discuss local issues, particularly related to management of forests. There are 12 office bearers, including a president and two executive members who are joint signatories of the association's bank account. At the time of writing, about MZM65 000 (US$2683) was available in this account, with more REDD payments (about US$22 942) expected to be transferred shortly. Decisions to use these funds are taken consensually by the members of the association. The fund has been used to support community development activities in the area, including the construction of two new school buildings in Nhambita and Bue Maria, and the provision of a primary health clinic in one of the villages.

Promotion of Microenterprises

In order to promote alternative livelihoods and improve local incomes, the project initiated and still supports several microenterprises (MEs), including the establishment of plant nurseries, the running of a community sawmill and carpentry shop, bee keeping and the setting up of a vegetable garden. The project has also hired many local people as agroforestry technicians, community extension workers, administrative staff and even as mechanics to repair project vehicles. In all, about 170 people are currently employed by various MEs and the project. Most of these people are drawn from the local community and are selected on the basis of their skill or aptitude for a certain activity. ME employees and project staff receive monthly wages. The average salary of people employed in the forest nursery is MZM1200 per month (US$50), while a senior agroforestry technician can receive as much as MZM15000 per month (US$619). Since these wages translate into additional cash flows into the community, any discussion of the impact of the project needs to differentiate between the impact of carbon payments and the impacts arising from the payment of these wages.

In the initial stages of the project, these MEs were supported by financial assistance from the EU. Since the end of this funding, these activities are being run as small businesses, with the aim of becoming self-sustaining. The carpentry shop, for example, receives orders from several local businesses including the hotel located within the GNP. However, many other enterprises such as the tree nursery still need financial support and are supported with some of the carbon revenues. In the long run, the project aims to turn even these activities into self-supporting businesses.

Research, Extension and Capacity Building

The Nhambita project has invested significant resources to develop computer models that can accurately measure changes in carbon stocks due to afforestation and reforestation activities, which have helped to provide an estimate of the deforestation rate in the project area (Grace et al. 2007). The hiring of extension workers under the project has helped to communicate the benefits of agroforestry in the local community and to train farmers in raising high value tree crops such as mango and cashew. Each extension worker supports about 100 farmers by visiting their fields and advising them on how to deal with pests, and so forth.

To date, the project has met most of its transaction and administrative costs from EU funding, which concluded in 2008. The project now has to meet these costs mainly from the revenue generated by the sale of carbon credits. Although each agroforestry contract is for 100 years, the project continues to produce additional offsets by expanding the number of farmers involved in the scheme and extending to new areas. For instance, since 2007, carbon agroforestry activities have been successfully introduced to the Zambézia and Cabo Delgado provinces in Mozambique. In the last few years, the project has successfully sold a total of 116 807 carbon offsets worth more than US$900 000. This is significant, considering that all of these offsets have been sold entirely through voluntary carbon markets, which absorb a smaller volume of carbon offsets than the Kyoto market. However, whether the project team can continue to raise such high revenue will depend on how markets change in response to the stance taken by the international community on the role of forestry in the United Nations Framework Convention on Climate Change.

IMPACTS OF THE NHAMBITA PROJECT

The impacts of the Nhambita project are discussed in terms of the level of environmental service created (that is, the total number of carbon offsets produced, with a discussion on issues of leakage and permanence), livelihood impacts on project participants and impacts on non-participating households. These impacts are explored using a mixed method strategy that combines quantitative and qualitative data. Quantitative data were gathered through two household surveys, one at the beginning of the project in June 2004 (the baseline survey) and the other at the completion of the pilot phase of the project in May 2008 (the follow-up survey).

In order to distinguish the impact of carbon payments from that of wages from employment in MEs, the follow-up survey was administered using a stratified random sampling technique. Based on the census conducted by Hegde and Bull (2008), the local population was divided into three strata, followed by random sampling of respondents within each stratum: (1) households both with an agroforestry or a carbon contract and with at least one member employed in an ME[7] (sample size n = 53); (2) households with only carbon sequestration contracts (n = 105); and (3) non-participating households that had neither carbon contracts nor employment in MEs (n = 47). The working hypothesis was that the positive impact of the project would be highest among those with both agroforestry contracts and employment in MEs, followed by households that possessed only agroforestry contracts and then non-participating households. We sampled about 20 per cent of potential respondents within each stratum; the total sample size of the follow-up survey was 205.

Qualitative data were collected through semi-structured discussions with an additional set of respondents: (1) respondents employed in various MEs and also contracted under agroforestry systems (group size 25); (2) women, most of whom had only agroforestry contracts (group size 25); (3) respondents who neither had carbon contracts nor employment in MEs (group size 14); (4) new immigrants to the area, most of whom did not possess carbon contracts (group size 24); and (5) members of the community association (group size 11). These data helped to interpret the quantitative data and picked up additional categories of local residents. Finally, field transects were conducted to understand the extent to which new land use practices were changing the local landscape.

Impact on the Environment

PES projects are distinguished from other subsidy-based conservation approaches by their stress on conditionality: payments are contingent on the level of provision of an environmental service (Wunder 2005). In this project, the level of provision was measured in terms of carbon offsets generated through carbon sequestration and REDD activities.

Carbon sequestration offsets

Agroforestry systems generate carbon sequestration offsets as per the following formula:

No. of carbon offsets (in tCO_2)[8] = {(kind of agroforestry contract − baseline) − buffer} × area in ha

The kind of agroforestry contract refers to one of the seven kinds of agroforestry systems offered, each with a specific carbon sequestration rate (see Table 8.1). The baseline refers to the existing carbon stock of vegetation (excluding food crops) on a site prior to new planting. To calculate the net number of carbon offsets, the baseline is subtracted from carbon sequestered through project activities. Since measuring actual carbon stock on each dispersed site is expensive, the project uses a set of sample plots to estimate the existing stock of carbon for the entire area (Sambane 2005).

Fifteen per cent of all carbon offsets are maintained as a buffer against risks of leakage and impermanence. Leakage refers to unplanned emissions of carbon arising from activities outside the project boundary. For instance, beneficiaries of a project may plant trees at one site but fell trees at another, resulting in a net release of carbon to the atmosphere. Impermanence pertains to the temporary nature of forestry carbon stocks: a forest can be cut at any stage, eventually releasing most of the sequestered carbon back into the atmosphere (Sedjo et al. 2001).

By May 2008, the project had 1234 agroforestry contracts covering about 1000 ha of land. Boundary planting was the most commonly adopted agroforestry system, accounting for 56 per cent of all contracts, followed by homestead planting (15 per cent) and fruit orchards (14 per cent). The project team estimated that, during the last 5 years, local farmers planted more than 500 000 trees under different agroforestry systems, generating a total of 82 056 tCO$_2$ as carbon sequestration offsets. Since these offsets have been calculated for sequestration rates over a 100-year period, they pertain to high value, long-term offsets as compared to low value temporary offsets that accrue from short-term mitigation activities (Haites 2004).

REDD offsets
Avoided deforestation or REDD activities are taken up in the form of protection and management of miombo woodlands around the GNP. On average, a well stocked area of miombo woodland contains 95.42 tCO$_2$ per ha (Grace et al. 2007). Analysis of remote-sensing images reveals that, between 1999 and 2007, the average rate of deforestation in miombo woodlands outside the GNP (an area of 48 952 ha in 1999) was 2.4 per cent per annum (Tipper 2008).[9] This rate of deforestation would denude the entire forest by 2040.

In order to reduce this deforestation rate, the Nhambita project pays the local community to protect 11 071 ha of miombo woodlands outside the GNP, while also motivating community members to conserve additional forest areas through selective logging and reforestation. The project team estimates that these protection and conservation activities have reduced emissions at the rate of 7.33 tCO$_2$/ha per annum since 2006, generating a total of 154 457 tCO$_2$ as REDD offsets.

In order to monitor compliance with protection contracts, the project team routinely makes surprise checks of miombo woodlands under REDD contracts. In addition, researchers from the University of Edinburgh carry out detailed investigations of remote-sensing images from the area to measure the fall in deforestation rates.

Leakage
The three main areas of leakage are tree-felling for charcoal production, uncontrolled burning of plots that are not covered under agroforestry contracts and clearance of forest for agriculture. Charcoal production is

an important source of livelihood in Mozambique (FAO 2007). Within Chicale *régulado*, charcoal production is mainly practised in the three villages of Mbulawa, Pungue and Povua along the national highway, and Herd (2007) estimates that 35 ha of local woodlands are lost every year due to tree-felling for charcoal production. The Nhambita project tries to tackle this issue by promoting agroforestry as an alternative source of income and by educating charcoal producers to use efficient kilns that reduce wood intake. The local community is also encouraged to resist locating any new charcoal kilns in the area.

Machambas are often burnt in preparation for cultivation, to clear undergrowth around settlements, for honey collection or to keep dangerous animals away. However, they also bring the risk of fire escaping to nearby forest areas resulting in significant carbon loss (Zolho 2005). Around 19 per cent of respondents in the follow-up survey confirmed that they had burned their *machamba* in the previous year, although most had modified their burning practice to reduce the risk of wildfire. The project also promotes tree species that improve soil productivity so that farmers have a stake in preserving the vegetation on their farms. Finally, scheduled carbon payments are withheld if contracted households are found to burn their *machambas* during the 7-year payment period for contracts.

The third and probably the most important driver of leakage in the area is the increase in the number of households and the resultant clearance of forest for agriculture. During the Mozambique civil war from 1977 to 1992, a large proportion of the population was displaced from rural areas to urban areas (Hatton et al. 2001). Since the return of peace in 1994, many of these people have returned to rural areas in search of livelihoods. For instance, 39 per cent of respondents in the follow-up survey had migrated to Chicale *régulado* from outside. This raises the question of whether the Nhambita project itself is attracting new immigrants. The 2008 survey indicates that this is not the case, as the three most cited reasons for the choice of Chicale *régulado* were availability of land to set up new *machambas*, relatives or family ties in the area and the presence of good forests. Dealing with internal migration is a national policy issue. Within the project context, migrants are encouraged to enrol for carbon contracts provided they agree to conserve existing forests and do not clear additional land for *machambas*.

Impact on the Livelihoods of Participants

At the time of the baseline survey (in 2004), most households in the area were extremely poor, with few sources of cash income. Table 8.2 illustrates that, by 2008, many of these statistics had improved.

Table 8.2   Change in selected statistics for households, 2004 and 2008

	Baseline survey (2004)	Follow up survey (2008)
Household size	4	5
Female headed household (%)	21	32
Number of literates per household	1	2
Households with no literates (%)	38	21
Number of *machambas* per household	3	2
Households cultivating food crops (%)	99	99
Households cultivating cash crop (%)	23	74
Households with at least one regular job (%)	9	32
Households involved in wage labour (%)	60	43
Number of households in sample	245	205

Source: Author's surveys in 2004 and 2008.

However, to what extent can these changes be ascribed to the Nhambita project as opposed to macro level changes in the economy? Pagiola et al. (2005) suggest an analytical approach that focuses on four key questions: (1) the extent to which poor people participate in the project; (2) the livelihood impacts on project participants; (3) the impacts on non-participants and wider spillover effects in the community; and (4) the extent to which impacts are due to causes other than the project in question.

In the context of the Nhambita project, where all households are eligible to enrol for agroforestry contracts, question 1 is answered by identifying factors that determine a household's decision to participate in the project. Questions 2 and 3 are answered using the sustainable rural livelihoods approach that considers both economic and non-economic impacts of a project on a local community (Landell-Mills and Porras 2002; Grieg-Gran et al. 2005; and outlined in the introduction to this volume). These impacts are considered at the household level, using the indicators outlined in Table 8.3, unless spillover effects at the community level are being discussed. Finally, question 4 helps to clearly state the limitations of the present approach while identifying ideas for future research. Important factors that may have influenced these results and how they can be addressed through further survey work in the community are also discussed.

Determinants of participation
Barriers to participation include tenure insecurity, non-availability of land, high transaction costs and high up-front investments needed to adopt new land-use practices (Grieg-Gran et al. 2005; Pagiola et al. 2005).

Table 8.3 Indicators to measure impact of Nhambita project on different kinds of capital

	Indicators
Financial capital	gross household income
	diversity in sources of income
	access to permanent job
	access to wage labour
	cultivation of commercial crops
	livestock ownership
	ownership of consumer durables
Human capital	gender division of labour
	average literacy rate
	investments in training and capacity building
	awareness of improved agricultural practices
Natural capital	land tenure and property rights
	permanence of agroforestry contracts
	perceived benefits from agroforestry practices
	status of food security (food purchased from outside)
Social capital	status of local leadership (community association)
	management of community trust fund
	level of women's participation in the project
Physical capital	status of education/health infrastructure
	access to market

The Nhambita project addresses some of these barriers; for instance, 30 per cent of the total carbon payment is provided during the first year to help meet the initial costs of planting trees.

Nevertheless, there could be additional factors that influence a household's decision to participate in the project. Based on econometric modelling, the key factors that had a significant influence on participation were: the gender of the head of the household, household size, total area of *machambas* owned by a family, the year the family migrated into the Chicale *régulado* and employment with an ME promoted by the Nhambita project. In contrast to many studies on agroforestry adoption, male-headed households had a lower probability of enrolling for agroforestry contracts than female-headed households. One possible explanation is that, due to their lower mobility, female-headed households find agroforestry contracts an attractive opportunity to earn cash income, in contrast to male-headed households that have other avenues, such as selling non-timber products from local forests (Hegde and Bull 2008).

The strongest determinant of participation was employment with a ME.

Indeed, almost all ME employees covered in the survey had agroforestry contracts. This could be due to their easy access to project staff as well as peer pressure from others who have already enrolled in the program. Household size and *machamba* area also had a positive effect on the likelihood of participation: the presence of additional household members helps to take care of increased labour requirements when a new land use practice is adopted. Similarly, a larger farm area enables a household to take some land out of crop production and devote it to trees. Finally, recent migrants to the community had a lower probability of participating than longer term residents. This could be due to their lack of familiarity with the community association, which often suggests names for inclusion in the project. Group discussion with new migrants confirmed this: many of them stated that they would like to participate in the project but their names were lower down the waiting list.

Other documented barriers to participation in carbon mitigation projects are transaction costs and tenure insecurity (Jindal et al. 2008). The project team estimates that 66 per cent of all carbon revenue from Nhambita will go towards meeting the costs of establishing, supervising, monitoring and marketing carbon contracts.[10] At present, these costs do not appear to influence participation rates in the area. In the initial stages of the project, most transaction costs were paid for by the donor, but subsequently they have been subtracted from the gross carbon revenue before payments are made to the community. Though this reduces the net payment to farmers, they remain higher than the opportunity cost of land. Transaction costs do, however, affect the sustainability of small-scale carbon sequestration projects. Where high transaction costs are unavoidable due to the participation of many small farmers, subsidies play a crucial role in the initial stages of the project. As subsidies dwindle, a reduction in net carbon payments to farmers occurs, which also decreases the poverty impact of household payments.

Under the new land act in the country, all land in the *régulado* is communally owned and individual households are allowed to take up subsistence farming on small plots of land (Zolho 2005). This allows the community members to retain the entire revenue from the sale of carbon offsets, particularly from the sale of REDD offsets on communally managed miombo woodlands. Individual tenure, however, is less secure as households do not possess title deeds. But once a household starts farming a certain plot of land, it has a de facto ownership of that plot. During our survey we found that almost all households owned *machambas* through this process and their participation in carbon sequestration activities was not restricted due to lack of access to land.

What do these results imply regarding participation of the poor in the

Nhambita project? Female-headed households tend to be poorer than male-headed households. Since they constitute a sizeable proportion of all households in the local community, it appears that poorer households are able to participate in the project. On the other hand, households with more resource endowments, in terms of farm area and employment in MEs, also have a higher probability of adopting agroforestry contracts. This seemingly contradictory evidence suggests that the project is equally accessible to both poor and relatively better-off households. Considering that almost 80 per cent of local households have enrolled for agroforestry systems and that more than 85 per cent of rural households in Sofala province were below the poverty line during the last decade (Simler et al. 2004), there is evidence that a significant number of poor households are participating in the Nhambita project.

Impact on financial capital

The Nhambita project provides two kinds of direct financial benefits to individuals: payments for adoption of agroforestry contracts that result in sequestration of carbon and monthly wages to people employed in various MEs. The schedule of agroforestry payments is as follows: 30 per cent of the contract value in the first year; 12 per cent per year for the next 5 years; and then a final payment of 10 per cent in the seventh year.[11] The average carbon payment received by the local community during 2007–08 was MZM1923 (US$79) per household. This was equivalent to about 2 months of wage labour.[12] During group discussions with project participants, many people said that they had used these carbon payments to buy roofing materials, food and clothes for the family, or books and school stationery for their children. Some people invested the money in better quality (agricultural) seeds, while others bought household durables such as radios or bicycles. For people employed in MEs, the average annual salary during the same period was MZM12484 (US$515).[13] Again, many respondents said that they had used this money to buy food for the family, or had invested the money in agriculture and in improving their houses.

Figure 8.2a indicates that improved access to a regular job is a major financial impact of the project. The majority of this increase was in households that possessed agroforestry contracts. Similarly, between 2004 and 2008, the proportion of local households that grew cash crops (sugarcane, cotton or tobacco) increased, and project participants recorded a slightly larger increase (75 per cent) than non-participating households (70 per cent). This indicates an increased availability of cash among participating households to pay for seeds and other inputs necessary to cultivate commercial crops. The diversion of land from food to cash crops could explain

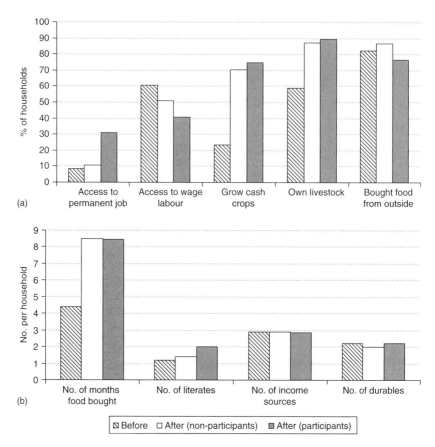

Notes: Figure 8.2a shows the proportion of sampled households for select indicators, while Figure 8.2b shows average value for sampled households on a different set of indicators. 'Before project' corresponds to year 2004 (n = 245), 'after project' to year 2008. After-project scenario is shown for two sub-groups: non-participating households (n = 47), and participating households (n = 158).

Figure 8.2   *Status of local households before (year 2004) and after*
             (2008) the project on selected indicators

why households now buy food for more months than before (Figure 8.2b). Since such changes in land use are reversible, this may not be a permanent trend. During the same period, the percentage of households that owned livestock also increased significantly, but there was a negligible difference in ownership patterns between project participants and non-participants. This makes it likely that the change was not entirely project related.

   In terms of income diversity (for example, sale of agricultural, animal or

Table 8.4   Results from one-way analysis of variance for cash income and ownership of selected durables among participating and non-participating households in 2008

	Non-participating households (Group 1)	Households that only hold agroforestry contracts (Group 2)	Households with both agro contracts and employment in micro-enterprises (Group 3)	F-statistic
Total mean annual cash income per household (MZN)	1200.40 (1425.75)	1435.40 (1950.10)	14,645.90 (5724.52)	319.41*
Mean number of durables per household	2.02 (1.31)	2.20 (1.44)	2.77 (1.12)	4.64*
Number of households in sample	47	105	53	

Notes: Figures in parentheses are standard deviations.
* Significant at 1%.

forest products, access to wage labour or a regular job), the range of income sources has remained almost the same at 2.9 per household. However, the number of income sources for project participants had decreased marginally to 2.85 by 2008, which supports Hegde and Bull's (2008) claim that project participants sell fewer forest products than others, this loss in forestry income being compensated by carbon payments.

In order to further distinguish the financial impact of carbon payments from that of wage employment in MEs, participants were split into two categories: households with agroforestry contracts only and households with both contracts and employment. Households from these two categories were then compared with non-participating households. One way analysis of variance was performed for two impact indicators: total annual cash income of a household in 2007–08, and average number of consumer durables.[14] The results in Table 8.4 show that the average annual cash income for households with both agroforestry contracts and employment in MEs (group 3) was much higher than either non-participating households (group 1) or households that hold only agroforestry contracts (group 2). Similarly, the average number of consumer durables owned was higher for group 3 than either for group 2 or group 1.

These results indicate that households with ME employment are better off than others, but the same does not apply to households that receive only carbon payments. While carbon payments do supplement household incomes, they are not, as yet, sufficient to have a significant financial impact at the household level. As more carbon payments are made over the next few years, it will be interesting to see whether this impact increases.

Impact on human capital
The impact on human capital is explored in terms of literacy achievement, training and capacity-building of local farmers, and changes in the gender division of labour. In 2004, the average number of literate people per household was 1.2. By 2008, this increased to 1.7 literate people per household, the increase being significantly higher in participating households than in non-participating households (see Figure 8.2b). In general, the proportion of households without a single literate member fell sharply from 38 per cent to 21 per cent. These improvements indicate two trends: (1) in general, literacy rates have increased in the area in line with macro-level improvements in educational attainment in Mozambique; and (2) project participants are relatively better positioned to access educational facilities in the area, as ascertained by sampled households, many of which used carbon payments to pay for school uniforms and stationery. In addition, the project has helped improve educational infrastructure in terms of setting up the community fund that paid for the construction of two new school buildings in the area.

Project staff also conduct regular training sessions for local farmers on improved agricultural practices. While it is difficult to quantify the impact of these training sessions, one indicator of their success is the adoption of 1234 agroforestry contracts and the establishment of more than 500000 new trees by local farmers within the last 5 years. In addition, all MEs promoted by the project are a source of on-the-job training and capacity-building for local residents. For instance, the carpentry and furniture shop has trained several people to be master carpenters and they now receive regular furniture orders from the hotel in the GNP.

One possible adverse impact of the project is an increase in women's work load. In the past, men and women jointly performed most agricultural operations, especially in male-headed households (Jindal 2004). However, since the onset of the project, women spend more time looking after newly planted trees. If their male counterparts are employed by the MEs, it means that women also perform most agricultural operations such as weeding, harvesting and setting up fire guards around their *machambas*.

This increase in work load was repeatedly raised during group discussions with women.

Impact on natural capital

According to the Land Act No.19/97, all land outside the GNP is managed by the local *régulo* as a common property resource of the entire community. The *régulo* can allot small pieces of land to community members for subsistence farming, but large-scale clearing of forest requires permission from the government. Since its inception in 2003, the Nhambita project has worked within this institutional framework and has, in fact, strengthened the efforts of the community to protect its natural resources. The project encourages participants to improve the management of existing *machambas* and dissuades them from clearing forest to open new *machambas*. Further, the community receives annual REDD payments for protecting nearly 11 000 ha of miombo forests from fire or any kind of harvesting. Although these payments provide a strong incentive for community members to preserve their natural capital, there are concerns regarding the sustainability of these efforts. With the high rate of migration into the community, there is an ever-increasing need to create new farms, resulting in loss of natural forest. It will be interesting to see what impact REDD activities have on deforestation rates in the long run, in the face of pressure to clear land for agriculture.

The permanence of carbon stocks generated through agroforestry on individual plots is questionable. The project generates high value or long-term carbon offsets, since they are estimated from sequestration activities over a period of 100 years. However, in order to provide an attractive return to farmers, all payments are made over the first 7 years. Although the project retains 15 per cent of income from all carbon offsets as a risk buffer, harvesting of trees in year 8 or anytime afterwards can jeopardize the viability of the entire project, as potential carbon buyers may shun the project. It is also unclear who will bear the risk of this potential impermanence. It would be difficult for the project to recover money from the farmers once they receive their payments, while buyers in voluntary markets rarely keep track of how permanent their carbon purchases are. On the other hand, if the project were to produce temporary carbon offsets by offering shorter contracts (say 10 years), the value of carbon offsets generated per hectare would be extremely low, reducing the payments to farmers. Permanence of carbon stocks is thus intrinsically tied to contract duration.

Finally, the sustainability of the project is affected by additional benefits from agroforestry. The project has introduced diverse tree species that include natives (for example, *Sclerocarya birrea*) as well as nitrogen-fixing trees (for example, *Gliricidia sepium*). Farmers can selectively cut these trees to obtain construction poles and firewood, and harvest other products such as honey and fodder. Inter-planting with nitrogen-fixing tree species can improve soil productivity. Many of these timber and non-timber benefits can increase household incomes. Since most of these indirect benefits are yet to materialize, they are not picked up in current indicators.

Impact on social capital

Social capital is reflected in the existence of trust among community members and the presence of strong local leadership. Like most rural areas in Mozambique, Chicale *régulado* is served by the traditional system of *régulo* and *mfumos*.[15] These local leaders are responsible for all development activities in the area. In addition, a democratically elected community association was formed by a local non-government organization in 2001 to help conserve natural resources in the area. These institutions have helped the project staff to penetrate deeper into the community and reach out even to remote settlements. The project, in turn, has expanded and strengthened these institutions. The community association, for instance, now helps the project staff with day-to-day management of the forest block, and in identifying households that wish to enrol for agroforestry contracts. The association is also responsible for managing the community trust fund that receives one-half of all REDD payments and a share of carbon sequestration payments for agroforestry contracts. These active forms of social capital have resulted in significant improvement in local infrastructure, some of which has been directly financed by the trust fund.

One weak aspect of the project is the absence of women in leadership positions. Though many women participate in the project, both as holders of agroforestry contracts and as employees of MEs, they play little role in the overall facilitation of the project. Usually women do not hold traditional leadership positions in the community, so they find it difficult to stake their claims to new spaces created by the project unless they are actively supported.

Impact on physical capital

When the carbon project was first introduced to Nhambita, public infrastructure was limited to a small primary school and two drinking water wells (Jindal 2004). Since then, there has been impressive growth: a newly constructed primary school, the opening of several new shops and a small primary health centre, and the development of several new wells for irrigation and drinking water. Given that similar developments have occurred in most other neighbouring villages that form the Chicale *régulado*, these are more a reflection of the general growth in Mozambique over the last few years than direct intervention by the Nhambita project.

The project has, however, contributed to an increase in disposable income of local households, which, in turn, generates demand for household items such as soap and cooking oil, and attracts investment in small businesses to provide such goods. On average, the project pays each contracted household about US$80 per annum in carbon payments. This translates into an aggregate payment of about US$70 000 per annum to participating households in the area. In addition, most people employed in MEs belong to the local community and spend a considerable proportion of their monthly wage within the area. As the project continues to scale up its activities, this cash inflow is bound to have a multiplier effect in terms of demand for infrastructure and other development activities in the area.

Impact on Livelihoods of Non-participants

By design, payments under a PES project are conditional on the provision of an environmental service by service providers (Wunder 2005). As a result, participating households are bound to gain more than non-participating households. However, there are two important concerns regarding non-participants: whether there are any systematic barriers that restrict their participation in the PES project and whether the PES project poses any risk to their livelihoods should they voluntarily decline to participate (Pagiola et al. 2005). The first of these concerns was discussed earlier.

Analysis of the second issue focuses on changes in opportunities to earn wage labour income in the area. Table 8.2 shows that prior to the project about 60 per cent of households earned wage labour income, but the 2008 survey found that this figure had dropped to 43 per cent. This decline was less pronounced for non-participating households (51 per cent) than for project participants (41 per cent), but still indicates a reduced demand for wage labour in the area. On the one hand, more households now cultivate cash crops which require intensive labour; on the other, as households adopt permanent vegetation cover in the form of agroforestry systems, they devote a smaller area to agricultural crops (food or cash crops), resulting in less demand for seasonal wage labour. For those households now employed in various MEs, the reduced access to wage labour is compensated. This is also true to a large extent for households that only receive carbon payments from agroforestry contracts. The same cannot be said for non-participating households, however. At this stage, there are insufficient data to establish whether or not this loss of wage labour income for non-participating households has resulted in an overall decline in their economic status. Comparison of average ownership of durables is also inconclusive in this regard (see Figure 8.2b). Further research is needed to examine how local farmers allocate *machamba* land to different crops including agroforestry, the impact of these decisions on demand for seasonal wage labour and the implications for economic status.

Limitations

A long-standing dilemma in socioeconomic studies is how to establish cause and effect. A project is part of a dynamic environment where many other elements co-exist. As such, ascribing a certain change to a project is a difficult exercise. A key variable that influenced our results is the impressive growth in Mozambique's economy over the past few years. Other variables include the increase in the staff strength of the GNP and the work of other development agencies in the area. We tried to deal with these variables through a three step process: (i) by identifying relevant changes in the local area by comparing the baseline scenario in 2004 with the status in 2008; (ii) by isolating macro level changes by disaggregating the sampled households into participants and non-participants (the assumption being that any large-scale phenomenon would uniformly affect all households in the area); and (iii) by further disaggregating participating households into those that were employed in MEs and those that only possessed agroforestry contracts to isolate the impact of carbon payments from other activities promoted by the project.

   Focus groups were also conducted in order to understand people's perceptions of the project and other changes to enable a richer interpretation of findings. For instance, discussion with the community association revealed that the GNP had a negligible impact on the local economy since most of the park staff were recruited from outside the local area. The study assumes that external changes have had a uniform effect on the community, but this assumption may not hold in all cases. For instance, participating households may have higher numbers of literate members not because of the project but because, in general, they are better able to access any externally funded development intervention. Another limitation is that this study relies heavily on a local census compiled by Hegde and Bull (2008). Though a nationwide census was completed recently in Mozambique, its results were not available at the time of the follow-up survey.

CONCLUSIONS: LESSONS FROM NHAMBITA

As negotiations continue on the role of forestry carbon projects in carbon mitigation strategies, this review highlights several relevant lessons. The combination of carbon sequestration activities on individual plots with REDD activities in community-owned forests presents an interesting option. This natural complementarity helps to reduce transaction costs. Establishing individual carbon contracts with smallholders and monitoring compliance with recommended land-use practices can be expensive. In contrast, community protection of forests can yield carbon offsets at a lower cost, especially when contiguous patches of forests are brought under REDD. Although transaction costs for the Nhambita project are high, they would have been much higher if the two activities were not combined. However, a word of caution must be expressed against bringing an entire area under carbon mitigation activities, leaving little flexibility for local communities to meet their timber and non-timber needs.

The menu of agroforestry systems offered by the Nhambita project also addresses the issue of flexibility to a certain extent. In contrast to many sequestration projects that allow only one set of land use practices, this menu provides flexibility for individual households to select systems that suit their specific needs. Mixing native trees with other multi-purpose species also ensures that, as the trees mature, farmers can fulfil many of their timber and non-timber requirements from their farmlands, reducing the need to fell forest trees.

This flexibility however, comes at a price: escalating transaction costs related to monitoring and supervising such individualized contracts. Even in Nhambita, where a large proportion of carbon offsets come from REDD activities, one-third of all carbon revenue is used to meet local transaction costs, and another third is paid to international brokers and commission agents who help in selling carbon offsets. As the experience of this project shows, meeting these transaction costs from carbon revenue alone is very difficult, especially for small-scale mitigation projects. This raises serious concerns about the viability of community-based carbon projects that are not subsidized by donor funds, at least in the initial stages of project development.

Another concern about the Nhambita project is the duration of the contract and the payment schedule. The idea of monetizing carbon offsets over 100 years and disbursing it over the first 7 years needs rethinking. Although this payment system takes care of farmer's up-front costs, it locks farmers into a very long-term contract, leaving little room for renegotiating the contract if market prices for carbon offsets increase in the future. Second, the issue of impermanence and the resultant liability need to be carefully evaluated. After the receipt of the last payment, there is a real risk that farmers, particularly future generations, may have little incentive to care for their trees. Instead of monetizing carbon offsets over 100 years, the project could pay farmers on an annual basis for the number of offsets they produce in that year. These payments would continue only as long as farmers protect their trees, thus giving them an incentive to continue caring for them.

In terms of livelihood impacts, this review finds that carbon payments do supplement household incomes, but it is too early to say whether or not beneficiary households will be able to move out of poverty permanently. As the trees become mature, products in terms of firewood, fruits, poles and other non-timber products will also play a crucial role in improving household incomes. It is also important to note that any similar project will face a choice between saturating fewer households with multiple contracts to achieve immediate economic impact or an equity-based approach of targeting more households with at least one contract, resulting in a smaller economic impact for each household. The Nhambita project has chosen the latter strategy, which may have a much more sustainable economic impact in the long run. As large tracts of forests usually exist in the form of common property resources in developing countries, this would be really pertinent for REDD where active participation of an entire community will be a prerequisite for successful adoption of any protection regime.

Finally, in terms of payment mechanisms for REDD projects, Nhambita demonstrates one option, where payments are distributed between wages for forest guards and a community fund. The judicious use of the fund becomes paramount in keeping alive the stakes of individual households in conserving the forest. However, forest use is dynamic and open to many conflicting claims. In the case of the Nhambita project, migration into the area and the need to create *machambas* for the new migrants places heavy pressure on forests, which cannot be managed with REDD payments alone. This is a national phenomenon which requires a countrywide strategy.

ACKNOWLEDGEMENTS

I am grateful to the University of Edinburgh, the Australian National University, and the Australian Agency for International Development (Australian Development Research Awards # EFCC 083) for providing me with a generous grant to undertake this study. Thanks to Sarah Carter, Joanne Pennie and Piet van Zyle for helping me at crucial stages. Acknowledgements to John Kerr, John Grace, Paul Ferraro, Robbie Richardson and two anonymous reviewers for their constructive feedback on earlier drafts of this study. I also express gratitude to members of the Nhambita community for their warm hospitality.

NOTES

1. The buffer zone is a designated area around the Gorongosa National Park, which is inhabited by local communities that were relocated there after the creation of the park.
2. *Régulado* is the traditional unit of land administration, managed by a local chief known as *régulo*. A *régulado* covers several thousand hectares of land and includes many villages or settlements.
3. International carbon markets deal in carbon offsets; each offset being equal to one ton of carbon dioxide sequestered by a forestry project.
4. Carbon offsets are sold to international buyers at a higher price. After deducting for its overheads and for commissions payable to brokers, the project pays farmers at the rate of US$4.50 per tCO_2. By comparison, in 2007 the average price of carbon on the Chicago Climate Exchange was US$3.13 per tCO_2 (Capoor and Ambrosi 2008).
5. This is similar to China's Grain for Green Programme where participating households receive economic incentives to take up new tree plantations for the first 5 years, but are expected to protect these trees for much longer (Uchida et al. 2005).
6. In September 2008, MZM 24.23 = US$1.
7. All households with a member employed in an ME also held at least one agroforestry contract.
8. Since carbon sequestration depends on the rate of biomass increase in a tree stand, sequestration is first measured in tons of carbon (tC) and then converted to tons of carbon dioxide (tCO_2).
9. Forest within the GNP has been well protected by the park management, resulting in negligible deforestation.
10. This is not an unusually high number for carbon projects; Michaelowa and Jotzo (2005) suggest that transaction costs can be as high as US$14.78 per tCO_2 for small-scale carbon projects.
11. REDD payments go directly into the community fund.
12. At the government prescribed rate of MZM51.15 per day, this comes to 38 days or about 1.5 months worth of wage labour. However, it is difficult for most people to find regular labour opportunities at this wage rate.
13. Some people, such as the forestry extension staff, receive a fixed monthly salary while others, such as carpenters, get a piece rate.
14. These included fishing rods, bicycles, radios, sewing machines, watches and cell phones, the list being modified from the 16 items used by Simler et al. (2004) in their assessment of poverty levels in Mozambique.
15. The *régulo* is the overall head of the *régulado*, *mfumos* serve under him and are responsible for managing smaller settlements within the *régulado*.

REFERENCES

Capoor, K. and P. Ambrosi (2008), *State and Trends of the Carbon Market 2008*, Washington DC: The World Bank.
Eraker, H. (2000), CO$_2$lonialism in Uganda, *NorWatch Newsletter* **5**, available at http://www.norwatch.no/index.php (accessed 16 November 2009).
FAO (2007), *State of the World's Forests: 2007*, Rome: Food and Agriculture Organization (FAO).
FAO (2003), Special Report: FAO/WFP Crop and Food Supply Assessment Mission to Mozambique, FAO Global Information and Early Warning System on Food and Agriculture, Food and Agriculture Organization and World Food Program.

Grace, J. (2008), *Miombo community land use and carbon management: Nhambita Pilot Project, Annual Report 1 August 2006–30 November 2007*, University of Edinburgh, UK, available at http://www.miombo.org.uk/Documents.html (accessed 16 November 2009).

Grace, J., C. Ryan and M. Williams (2007), An inventory of tree species and carbon stocks for the Nhambita Pilot Project, Sofala Province, Mozambique, University of Edinburgh, UK, available at http://www.miombo.org.uk/Documents.html (accessed 16 November 2009).

Grieg-Gran, M., I. Porras and S. Wunder (2005), How can market mechanisms for forest environmental services help the poor? Preliminary lessons from Latin America, *World Development* **33** (9), 1511–27.

Haites, E. (2004), Rewarding sinks projects under the CDM, *Environmental Finance* March 2004.

Hatton, J., M. Couto and J. Oglethorpe (2001), *Biodiversity and War: A Case Study of Mozambique*, Washington DC: Biodiversity Support Program.

Hegde, R. and G. Bull (2008), Economic shocks and Miombo woodland resource use: a household level study in Mozambique, Department of Forest Resource Management, University of British Columbia.

Heltberg, R., K. Simler and F. Tarp (2003), Public spending and poverty in Mozambique, Food Consumption and Nutrition Division Discussion Paper No. 167, Washington DC: International Food Policy Research Institute.

Herd, A. (2007), Exploring the socio-economic role of charcoal production and the potential for sustainable production in the Chicale Régulado Mozambique, MSc dissertation, School of GeoSciences, University of Edinburgh.

Jindal, R. (2004), Measuring the socio-economic impact of carbon sequestration on local communities: an assessment study with specific reference to the Nhambita pilot project in Mozambique, MSc dissertation, School of GeoSciences, University of Edinburgh, UK.

Jindal, R., B. Swallow and J. Kerr (2008), Forestry-based carbon sequestration projects in Africa: potential benefits and challenges, *Natural Resources Forum* **32**, 116–30.

Jindal, R. and J. Kerr (2007), *USAID PES Sourcebook: Lessons and Best Practices for Pro-poor Payment for Ecosystem Services*, Blacksburg, Virginia: Office of International Research, Education, and Development (OIRED), available at http://www.oired.vt.edu/sanremcrsp/menu_research/PES.Sourcebook. Contents.php (accessed 16 November 2009).

Landell-Mills, N. and I.T. Porras (2002), *Silver Bullet or Fool's Gold? A Global Review of markets for forest environmental services and their impact on the poor*, London: International Institute for Environment and Development.

Michaelowa, A. and F. Jotzo (2005), Transaction costs, institutional rigidities and the size of the clean development mechanism, *Energy Policy* **33**, 511–23.

Miranda, M., I.T. Porras and M.L. Moreno, (2003), *The Social Impacts of Payments for Environmental Services in Costa Rica: A Quantitative Field Survey and Analysis of the Virilla Watershed*, UK: International Institute for Environment and Development.

Pagiola, S., A. Arcenas and G. Platais (2005), Can payments for environmental services help reduce poverty? An exploration of the issues and the evidence to date from Latin America', *World Development* **33** (2), 237–53.

Sambane, E. (2005), Above ground biomass accumulation in fallow fields at the Nhambita Community, Mozambique, MSc dissertation, University of Edinburgh, UK.

Sedjo, R., G. Marland and K. Fruit (2001), *Renting Carbon Offsets: The Question of Permanence*, Washington DC: Resources for the Future.

Simler, K.R., S. Mukherjee, G.L. Dava and G. Datt (2004), *Rebuilding after War: Microlevel Determinants of Poverty Reduction in Mozambique*, Research Report 132, Washington DC: International Food Policy Research Institute (IFPRI),.

Tipper, R. (2008), *Template for Plan Vivo Technical Specifications on Avoided Deforestation Conservation of miombo woodland in central Mozambique*, Edinburgh Centre for Carbon Management, UK, available at http://www.miombo.org.uk/Documents.html (accessed 16 November 2009).

Tipper, R. (2002), Helping indigenous farmers to participate in the international market for carbon services: the case of Scolel Te, in S. Pagiola, J. Bishop and N. Landell-Mills (eds), *Selling Forest Environmental Services: Market-based Mechanisms for Conservation and Development*, London: Earthscan, pp. 223–34.

Uchida, E., J. Xu and S. Rozelle (2005), Grain for green: cost-effectiveness and sustainability of China's conservation set-aside program, *Land Economics* **81** (2), 247–64.

Uchida, E., J. Xu, Z. Xu and S. Rozelle (2007), Are the poor benefiting from China's land conservation program? *Environment and Development Economics*, **12**, 593–620.

University of Edinburgh (2002), Nhambita Pilot Project: Project Proposal, Project proposal submitted by the University of Edinburgh to the European Commission.

Williams M., C.M. Ryan, R.M. Rees, E. Sambane, J. Fernando and J. Grace (2008), Carbon sequestration and biodiversity of re-growing miombo woodlands in Mozambique, *Forest Ecology and Management* **254**, 145–55.

World Bank (2008), *World Development Report 2008: Agriculture for Development*, Washington DC: The World Bank.

Wunder, S. (2005), Payments for environmental services: some nuts and bolts, CIFOR Occasional Paper No. 42, Bogor, Indonesia: Centre for International Forestry Research.

Zolho, R. (2005), Effect of fire frequency on the regeneration of Miombo Woodland in Nhambita, Mozambique, MSc thesis, University of Edinburgh, UK.

9. Poor household participation in payments for environmental services in Nicaragua and Colombia

Ana R. Rios and Stefano Pagiola

INTRODUCTION

The Regional Integrated Silvopastoral Ecosystem Management Project, implemented at sites in Colombia, Nicaragua and Costa Rica from 2003 to 2008, offers an excellent opportunity to examine the ability of poor households to participate in Payments for Environmental Services (PES). The Silvopastoral Project used PES to stimulate the adoption of silvopastoral practices in degraded pastures by paying participating households for the biodiversity conservation and carbon sequestration services that were generated, with financing from the Global Environment Facility (GEF). Unlike many other PES schemes, the Silvopastoral Project offered a wide range of participation options, ranging from simple and inexpensive land-use changes to substantial and complex changes (with correspondingly higher payments). That some of the choices offered by the project are complex and onerous provides a particularly strong test of poorer households' ability to participate.

In this chapter, we evaluate the extent to which poor households were able to participate in the Silvopastoral Project's PES scheme, using data from two of its sites. As the same payment scheme was offered in both areas, we are able to compare poor household's participation in PES under different agronomic and socioeconomic conditions. In particular, one site is characterized by high levels of poverty, with most households falling below the poverty line, and many below the extreme poverty line, while the other site exhibits a very wide range of income levels, including both extremely poor and very wealthy farm households. Because of the nature of the practices being promoted, the Silvopastoral Project's welfare impact cannot be assessed at this time. What is assessed is the threshold issue of ability to participate: if poorer households cannot participate, they will not receive any benefits, whether large or small.

CONSTRAINTS TO THE PARTICIPATION OF POOR HOUSEHOLDS IN PES SCHEMES

PES schemes pay land users to maintain or switch to land uses that provide environmental services that others value (Wunder 2005; Pagiola and Platais 2007; Engel et al. 2008). Participation is voluntary, and participants receive payments for doing so. This creates a prima facie presumption that participants are at least no worse off by joining than they would be by not joining. Were this not the case, they could simply decline to participate.

PES represent a potential additional income source for land users, but these potential benefits will only be realized by those who participate. However, many observers have feared that poorer households would be unable to participate in PES schemes. Pagiola et al. (2005) identified three categories of factors that might affect a household's participation in a PES scheme: factors that affect the eligibility to participate; factors that affect the desire to participate; and factors that affect the ability to participate. The three categories form a logical sequence: the ability to participate only becomes an issue for households that wish to do so, and that, in turn, is only relevant for households that are eligible to participate.[1]

Eligibility to participate is affected by the scheme's targeting and by requirements it may impose, which may affect poorer households differentially. The eligibility of poorer households is not an issue in this study, however, as it focuses on areas that were already selected for inclusion in the project.

Assuming that a given household is eligible to participate, whether it desires to participate is likely to depend primarily on whether it expects to be better off as a result. Participation in PES can be thought of as adopting a particular production technique, whose returns include payments from the scheme. The literature on technology adoption and programme participation thus provides many insights into the factors likely to affect participation (Feder et al. 1985). The literature on adoption of agroforestry practices (Pattanayak et al. 2003; Mercer 2004) is particularly pertinent here, as the practices promoted at the study sites are very similar. Previous analyses confirm the significance of factors that tend to affect the benefits or the costs of participation, such as prices faced, farm characteristics, the opportunity cost of household labour, the fit in the farming system or the risk involved (Pattanayak et al. 2003). The slope of the land, for example, can affect the extent to which productivity is threatened under current practices, thus increasing incentives to adopt land uses that are less vulnerable to degradation. As developing-country PES schemes typically offer fixed payments per hectare for adopting a given practice, the payment amount is unlikely to differentially affect the desirability of participation across households.

A household may want to participate in a PES scheme and yet be unable to do so for a variety of reasons. Participation in a PES scheme requires adoption of the land uses promoted by the scheme. This may be simple and cheap, if the scheme calls for retaining existing land uses (as in the Costa Rica scheme's forest protection contract), or it may be complex and costly, if the scheme calls for switching to new practices (as in the PES scheme studied here). Tenure issues are often critical, particularly when PES schemes require long-term investments, such as reforestation or adoption of silvopastoral practices. Tenure variables were significant in 72 per cent of agroforestry adoption studies that included them, with greater security of tenure being consistently associated with greater adoption (Pattanayak et al. 2003). In Costa Rica, Thacher et al. (1997) and Zbinden and Lee (2005) found tenure-related variables to be highly significant in explaining participation in the country's PES scheme and its predecessors. When the practices to be adopted are complex, access to technical assistance (TA) may be an issue. Access to extension was found to significantly affect agroforestry adoption in 90 per cent of studies that included it (Pattanayak et al. 2003), including two studies in Costa Rica (Thacher et al. 1997; Zbinden and Lee 2005). Adopting new land-use practices may also prove difficult if households cannot finance the necessary investments. Assets and credit both tend to increase adoption of agroforestry practices, and their role is very often significant (Pattanayak et al. 2003).

Many factors that affect a household's ability to participate in PES are likely to be more salient for poor households. Poorer households are less likely to have secure tenure, tend to have fewer savings and less access to credit, and are less likely to receive TA (de Janvry and Sadoulet 2000; López and Valdés 2000). Whether poor households are able to participate in PES schemes (assuming that they are eligible and interested in doing so) is thus a legitimate concern.

THE SILVOPASTORAL PROJECT

The Regional Integrated Silvopastoral Ecosystem Management Project piloted the use of PES in three sites in Colombia, Costa Rica and Nicaragua (Pagiola et al. 2004). The project was financed by a US$4.5 million grant from the GEF, through the World Bank. It was implemented in the field by local non-government organizations (NGOs).

PES Scheme Design

The Silvopastoral Project used PES to promote the adoption of silvopastoral practices in areas of degraded pastures, with the aim of generating biodiversity conservation and carbon sequestration services. Silvopastoral practices include: (1) planting high densities of trees and shrubs in pastures; (2) cut and carry systems, in which livestock is fed with the foliage of specifically planted trees and shrubs ('fodder banks'); and (3) using fast-growing trees and shrubs for fencing and wind screens. These practices provide deeply rooting, perennial vegetation that has a dense but uneven canopy.

The on-site benefits of silvopastoral practices to land users may include additional production from the trees, such as fruit, fuelwood, fodder or timber; maintaining or improving pasture productivity by increasing nutrient recycling; and diversification of production (Dagang and Nair 2003). These benefits can be important, but are often insufficient to justify adopting silvopastoral practices – particularly practices with substantial tree components, which have high up-front planting costs and only bring benefits several years later.

Because of their increased complexity relative to traditional pastures, silvopastoral practices also have important biodiversity benefits (Dennis et al. 1996; Harvey and Haber 1999). They can also fix significant amounts of carbon in the soil and in the standing tree biomass (Fisher et al. 1994; Swallow et al. 2007). Both biodiversity and carbon sequestration benefits are off-site, however, so land users tend not to include them in their decisions about which practices to adopt. GEF funding for the Silvopastoral Project was based on the desire to secure these biodiversity and carbon sequestration benefits.[2] Silvopastoral practices can also affect water services, though the specific impact is likely to be site-specific (Bruijnzeel 2004). The Silvopastoral Project did not pay for water services.

To encourage the adoption of more beneficial practices, the Silvopastoral Project offered payments that were proportional to the level of services provided. To do so, it developed indices of the biodiversity conservation and carbon sequestration services provided by different land uses, then aggregated them into an 'environmental services index' (ESI).[3] The project distinguished 28 different land uses, each with its own ESI score, and paid participants according to the change in total ESI score over their entire farm area. Remote-sensing imagery, followed by on-the-ground verification was used to prepare detailed baseline land-use maps of each PES recipient household (as well as of members of a control group, see Appendix 1), indicating the land use being undertaken on each parcel. Land-use changes on each parcel were then monitored annually and payments were made based on the observed land-use changes. Deforestation was prohibited, as was the use of fire as a management tool; either would result in a halt to further payments.

Silvopastoral practices tend to be unattractive to land users, despite their long-term benefits, primarily because of their substantial initial investment, and the time lag between investment and returns. This led to the hypothesis that a relatively small payment provided early on could 'tip the balance' between current and silvopastoral practices, by increasing the net present value of investments and reducing the period in which these practices impose net costs on land users (Pagiola et al. 2004; 2008). On this basis, participating households received payments of US$75 per incremental ESI point per year,[4] over a 4-year period.[5] Payments were made annually, after on-the-ground verification of land-use changes. They also received a one-time payment of US$10/point for the baseline points.[6]

The Silvopastoral Project offers a strong test of the ability of poorer households to participate in PES, as many of the measures it supported are both expensive and technically challenging to implement. At the same time, the project also offers several easier and cheaper options. Should either investment requirements or technical capacity prove to be insurmountable obstacles for poorer households, there would be a clear division in the PES-supported activities they implement.

Study Sites

The Silvopastoral Project's sites were in Quindío, Colombia; Esparza, Costa Rica; and Matiguás–Río Blanco, Nicaragua (Pagiola et al. 2004). In this chapter, we focus exclusively on the Quindío and Matiguás–Río Blanco sites. (Appendix 9.1 describes the data sources used.)

The Matiguás–Río Blanco area is located in the department of Matagalpa, about 140 km northeast of Managua, on the southern slopes of the Cordillera de Darien. It has an undulating terrain, with an elevation of about 300–500 metres above sea level. Average temperature is about 25°C and average rainfall is between 1700 mm and 2500 mm per year. Farms range in size from 10–30 ha, with a few over 60 ha. Most households are poor, with many falling below Nicaragua's extreme poverty line.

Land use in Matiguás–Río Blanco is dominated by extensive grazing. As shown in Table 9.1, pastures accounted for about 63 per cent of the area before the start of the project. Of this, about half was degraded pasture, and a little over a quarter had either no or few trees. Annual crops made up a very small part of the total area. About 20 per cent of the total area remained under forest, mostly along streambanks. Silvopastoral practices, though not common, were not unknown even before the project: there

Table 9.1   Land use among Silvopastoral Project PES recipients, Matiguás-Río Blanco, Nicaragua

Land use	Environ-mental services index (points/ha)	Before project (2003)		Year 4 of project (2007)	
		(ha)	(%)	(ha)	(%)
Infrastructure, housing, and roads	0.0	11.4	0.4	14.8	0.5
Annual crops	0.0	212.0	7.6	78.6	2.8
Degraded pasture	0.0	780.6	28.0	147.6	5.3
Natural pasture without trees	0.2	43.0	1.5	27.9	1.0
Improved pasture without trees	0.5	22.6	0.8	25.6	0.9
Semi-permanent crops	0.5	37.0	1.3	18.8	0.7
Natural pasture with low tree density	0.6	279.3	10.0	231.5	8.3
Fodder bank[a]	0.8	79.8	2.9	238.9	8.6
Improved pasture with low tree density	0.9	134.4	4.8	213.6	7.7
Natural pasture with high tree density[b]	1.0	338.1	12.1	522.4	18.7
Diversified fruit crops[a]	1.1	18.5	0.7	23.8	0.9
Monoculture timber plantation	1.2	1.1	0.0	4.7	0.2
Improved pasture with high tree density[b]	1.3	151.3	5.4	537.8	19.3
Scrub habitats (*tacotales*)	1.4	137.7	4.9	134.3	4.8
Secondary and riparian forest[a]	1.7	543.2	19.5	569.8	20.4
Total area		2790.0	100.0	2790.0	100.0
Live fence (km)	1.1	115.8		325.5	

Notes: Totals may not correspond exactly to data because of rounding.
Land uses recognized by the project but not found at this site are omitted.
Includes land use by PES recipients only.
[a]   Similar land uses with small areas have been aggregated; ESI shown is for use with largest area.
[b]   The project distinguishes land uses with recently planted trees from the same land uses with mature trees for the purpose of computing the ESI score; here these land uses have been aggregated to their mature state and the corresponding ESI score is shown.

Source: ESI score from Silvopastoral Project manual (CIPAV 2003); land use from Silvopastoral Project, based on analysis of remote-sensing imagery verified in the field

were some 489 ha of pastures with high tree density and 88 ha of fodder banks, for example.

The Quindío area is located in Colombia's Central Cordillera, in the watershed of Río La Vieja, at an altitude of about 900–1500 metres above sea level. Average temperature is about 20–25°C and average annual rainfall between 1500 mm and 2000 mm. Farms range from 10–20 ha to some of 50–80 ha. In this former coffee area, many of the larger farms are owned by urban professionals and managed by employees (*mayordomos*).

Extensive grazing was the main land use in Quindío prior to project start (see Table 9.2). Coffee was once dominant, but was replaced by pasture during the last decade due to low coffee prices and now accounts for less than 1 per cent of the area. Degraded and treeless pastures dominated the landscape, accounting for about 65 per cent of the area. Livestock production is primarily for meat, with a small proportion being used for milk production. Overall tree cover was low, although there were significant forest remnants, mostly riparian forest. Silvopastoral practices were practically non-existent. Only seven in 110 farms surveyed had fodder banks, for example, with an average of less than 1 ha each.

Neither Nicaragua nor Colombia is among the top deforesters, but both have high deforestation rates. Conditions at Matiguás–Río Blanco are broadly representative of those observed in many parts of Central America, while conditions in Quindío are broadly representative of conditions found in parts of Venezuela, Brazil and Ecuador.

All households in the study sites that owned livestock were eligible to participate. Participants were selected on a first-come basis until the available funding was exhausted. The characteristics of participating households are summarized in Tables 9.3 and 9.4. At Matiguás–Río Blanco, participating households had six members on average, about 34 ha of land, and about 23 livestock, of which a third were cows.[7] The average annual per capita income of about Nicaraguan Cordoba (C\$) 2000 (US\$140) was below the national poverty line.[8] Other indicators confirm the low living standards of the area's households: few had water or electricity, and education levels were very low. Agriculture was the main economic activity, with few households having off-farm income. In Quindío, the average household had fewer than five members, about 36 ha of land, and a herd of about 57 livestock units. Average annual per capita income was about Colombian Pesos (COP) 10 million (US\$3700), but with very high variation across the sample.

To assess relative participation levels, households at each site were classified into three groups based on their estimated income.[9] In Matiguás–Río Blanco, the national poverty line for 2001 (World Bank 2003), adjusted for inflation, was used to divide households into those with incomes below the extreme poverty line ('extremely poor'), those with incomes between the extreme poverty line and the poverty line ('poor') and those with incomes above the poverty line ('non-poor'). In Quindío, with its much greater spread of income levels, there was a clear jump in income

Table 9.2   Land use among Silvopastoral Project PES recipients, Quindio, Colombia

Land use	Environ-mental services index (points/ha)	Before project (2003)		Year 4 of project (2007)	
		(ha)	(%)	(ha)	(%)
Annual crops	0.0	37.9	1.3	37.2	1.3
Degraded pasture	0.0	78.3	2.7	7.1	0.2
Natural pasture without trees	0.2	721.5	24.9	239.5	8.3
Improved pasture without trees	0.5	1078.8	37.3	873.0	30.2
Semi-permanent crops (plantain, sun coffee)	0.5	184.1	6.4	148.4	5.1
Natural pasture with low tree density	0.6	6.2	0.2	10.4	0.4
Diversified fruit crops	0.7	73.7	1.9	59.7	2.1
Fodder banks[a]	0.8	4.6	0.0	27.5	1.0
Improved pasture with low tree density	0.9	54.8	2.5	333.4	11.5
Natural pasture with high tree density[b]	1.0	0.0	0.2	67.9	2.3
Shade-grown coffee	1.3	23.5	0.8	33.8	1.2
Improved pasture with high tree density[b]	1.3	2.2	0.1	266.5	9.2
Bamboo (*guadua*) forest	1.3	43.9	1.5	52.6	1.8
Timber plantation[a]	1.4	0.0	0.0	5.5	0.2
Scrub habitat (*tacotales*)	1.4	48.8	1.7	42.0	1.5
Riparian forest	1.5	369.2	12.8	392.8	13.6
Intensive silvopastoral system	1.6	0.0	0.0	130.2	4.5
Primary and secondary forest[a]	2.0	165.7	5.7	165.7	5.7
Total area		2893.2	100.0	2893.2	100.0
Recently established live fence (km)	0.6	1.4		255.5	
Multistory live fence or wind break (km)	1.1	0.7		92.9	

Notes: Totals may not correspond exactly to data because of rounding.
Land uses recognized by the project but not found at this site are omitted.
Includes land use by PES recipients only.
[a]   Similar land uses with small areas have been aggregated; ESI shown is for use with largest area.
[b]   The project distinguishes land uses with recently planted trees from the same land uses with mature trees for the purpose of computing the ESI score; here these land uses have been aggregated to their mature state and the corresponding ESI score is shown.

Source: ESI score from Silvopastoral Project manual (CIPAV 2003); land use from Silvopastoral Project, based on analysis of remote-sensing imagery verified in the field

Table 9.3   Characteristics of participating households, Matiguás-Río Blanco, Nicaragua

Variable	PES recipients				Control group	Entire sample
	Extremely poor	Poor	Non-poor	All		
Income per capita ('000 C$)	−2.9[ab]	4.2[ac]	15.8[bc]	4.5	−7.5	2.0
Assets ('000 C$)	7.3	11.7	2.6	6.8	21.4	9.8
Farm area (ha)	23.7[b]	30.9	40.1[b]	30.3[d]	48.2[d]	34.0
Cattle (livestock units)	13.7[b]	14.0[c]	32.8[bc]	19.8[d]	36.8[d]	23.3
Hilly topography (% farm area)	16.2	15.0	24.9	18.7	23.0	19.6
Water (% with water service)	20.5[b]	21.1	41.4[b]	27.2	25.0	26.7
Electricity (% with electric service)	2.3[b]	0.0[c]	17.2[bc]	6.5	16.7	8.6
Access by road all year round (%)	79.5[b]	84.2	96.6[b]	85.9[d]	100.0[d]	88.8
Paved road (%)	6.8	10.5	20.7	12.0	20.8	13.8
Family labour (hours/ha/ week)	5.4[ab]	3.1[a]	3.1[b]	4.2	3.4	4.0
Household size (members)	7.7[ab]	6.2[ac]	4.8[bc]	6.4[d]	5.1[d]	6.2
Dependency ratio (children per adult)	1.0[ab]	0.6[a]	0.6[b]	0.8	0.7	0.8
Experience (years)	10.0	13.7	13.3	11.8	10.0	11.4
Education of household head (years)	2.3	3.4	3.1	2.8	3.8	3.0
Male headed household (%)	86.4	94.7	96.6	91.3	87.5	90.5
Off-farm work (% with off-farm employment)	22.7	26.3	10.3	19.6[d]	4.2[d]	16.4
Off-farm income (% of total income)	−0.8	0.1	0.0	−0.4	0.0	−0.3
Non-farm enterprise (% owners)	18.2	10.5	24.1	18.5[d]	4.2[d]	15.5
Technical assistance (% with current access)	25.0	31.6	31.0	28.3[d]	12.5[d]	25.0
Number of observations	44	19	29	92	24	116

Notes: [a,b,c,d], indicate means are significantly different in paired *t*-test at 10 per cent test level. (If two figures in a row both have the superscript 'ᵃ', the difference between them is statistically significant; if they do not share a superscript, the difference between them is not statistically significant.)
Extremely poor <C$2943 (US$210); poor ≥C$2943 (US$210), <C$5639 (US$400); non-poor ≥C$5639 (US$400).
Children are household members under 12.

Table 9.4   Characteristics of participating households, Quindío, Colombia

Variable	PES recipients				Control group	Entire sample
	Low income	Middle income	High income	All		
Income per capita (million COP)	-0.7^{ab}	7.0^{ac}	39.9^{bc}	8.2	14.3	10.0
Assets (million COP)	4.8^{ab}	8.8^{a}	17.9^{b}	8.4	8.7	8.5
Farm area (ha)	23.3^{ab}	49.2^{a}	62.1^{b}	40.2^{d}	25.4^{d}	36.0
Cattle (livestock units)	44.3^{a}	77.4^{b}	184.2^{ab}	60.1	48.5	56.8
Flat (% farm area)	19.1	25.9	24.5	22.9^{d}	36.9^{d}	26.9
Distance to nearest village (km)	7.9^{a}	7.2	4.3^{a}	7.1^{d}	5.24^{d}	6.6
Water (% with water service)	90.0^{a}	96.9	100.0^{a}	94.4	96.6	95.0
Farm resident (%)	36.7	22.0	40.0	30.6	17.2	26.7
Family labour (man-days/ha/yr)	11.1^{a}	7.4^{b}	3.1^{ab}	8.3	nd	nd
Household size (members)	5.2^{a}	4.9^{b}	3.6^{ab}	4.9^{d}	3.7^{d}	4.5
Dependency ratio (children per adult)	0.25^{a}	0.55^{a}	0.36	0.40^{d}	0.22^{d}	0.35
Age of household head (years)	46.8^{ab}	40.7^{a}	38.7^{b}	42.9	43.9	43.2
Literacy of household head (%)	96.7	93.8	100.0	95.8	93.1	95.1
Education of household head (years)	4.0^{a}	5.2	8.6^{a}	5.2	4.3	4.9
Off-farm work (% with off-farm employment)	10.0	15.6	20.0	13.9	10.3	12.9
Technical assistance (% with current access)	33.3	34.4	50.0	36.1^{d}	10.3^{d}	28.7
Credit (% with access to credit)	23.3	25.0	40.0	26.4	13.8	22.8
Number of observations	30	32	10	72	29	101

Notes: ^{a,b,c,d} indicate means are significantly different in paired *t*-test at 10 per cent test level. (If two figures in a row both have the superscript 'a', the difference between them is statistically significant; if they do not share a superscript, the difference between them is not statistically significant.)
nd = no data.
Low income < COP2 million (US$750); middle income ≥ COP2 million (US$750), < COP20 million (US$7500); high income ≥ COP20 million (US$7500).
Children are household members under 12.

levels above COP20 million (US$7500), so households with income above this level were grouped into a 'high-income' group. Below this income level, there was no apparent clustering, so the remaining households were divided into two groups of similar size, with the division falling at COP2 million (US$750). Households with incomes between COP2 million and COP20 million (US$749–7500) were placed in a 'middle-income' group, and those with incomes below COP2 million (US4750) were placed in a 'low-income' group. About half of the latter group falls below Colombia's official poverty line (World Bank 2002).

At both sites, poorer households have significantly less land and smaller herds than better-off households. They also have larger households and more dependents per adult, although in Quindío the proportion of dependents is highest among middle income households. In Matiguás–Río Blanco, differences in educational level and experience are minimal and not statistically significant; these differences are much greater in Quindío, but are also not statistically significant. In Matiguás–Río Blanco, average access to services is low, but poorer households are less likely to have either electricity or water; in contrast, average access to services is high in Quindío, but low income households are less likely to have water services and are more likely to live further from the nearest village. Particularly important for what follows, access to credit and TA is highest among better-off households at both sites, although the differences are not significant. The topography of farms is broadly similar across income groups at both sites.

PES Implementation

The Silvopastoral Project paid for baseline ESI points at both sites in July 2003. It made its first payments for changes in land use in May 2004. Additional payments were made in 2005, 2006 and 2007.

Tables 9.1 and 9.2 compare land use by PES recipients before the project and after four years of payments. Overall, there was substantial land-use change at both sites. Some form of land-use change was found on 1343 ha (48 per cent of total area) in Matiguás–Río Blanco, and on 1258 ha (44 per cent of total area) in Quindío.[10] Changes ranged from minor (such as sowing improved grasses in degraded pastures) to very substantial (such as planting high-density tree stands or establishing fodder banks). The area of degraded pasture fell by over 80 per cent in Matiguás–Río Blanco and by over 90 per cent in Quindío. The area of natural pastures without trees declined by a third in Matiguás–Río Blanco and by two-thirds in Quindío. In Matiguás–Río Blanco, the greatest increase was in the area of pasture with high tree density, which increased by 570 ha (116 per cent). The area of fodder banks increased by almost 160 ha (200 per cent), and about 210 km of live fencing were established. In Quindío, most of the gains were experienced in pastures with high tree density, which increased from almost nothing to 334 ha. The area of fodder banks increased relatively little (from less than 5 ha to over 28 ha), but that of intensive silvopastoral systems (*Leucaena* planted at 5000 trees/ha) increased substantially (from 0 ha to 130 ha). About 346 km of live fencing were established. These changes increased the total ESI score of PES recipients by 53 per cent in Matiguás–Río Blanco and by 49 per cent in Quindío; households in Matiguás–Río Blanco made more substantial changes on a smaller proportion of their land than households in Quindío.

In Quindío, the land-use changes undertaken by PES recipients were vastly greater than those observed in the control group – less than 13 per cent of the control group's land experienced any change, for an increase of only 7 per cent in ESI points. The lack of a proper control group at Matiguás–Río Blanco prevented a similar comparison, but casual observation suggests that land-use changes among non-recipients in nearby areas were substantially less extensive, in both area affected and degree of change.

POOR HOUSEHOLD PARTICIPATION IN THE SILVOPASTORAL PROJECT'S PES SCHEME

The question of interest here is the extent to which poorer households were able to participate in this success. Figures 9.1 and 9.2 break down observed land-use changes by household income group. At Matiguás–Río Blanco, poor and extremely poor households accounted for 49 per cent of the decline in degraded pasture. In Quindío, low income households accounted for 35 per cent of the decline in degraded pastures and 45 per cent of the decline in improved pasture without trees. They accounted for only 9 per cent of the decline in natural pasture without trees, but this is primarily due to their having the least area in this category of any of the income groups.

Land-use changes by poorer households at both sites were not limited to adopting technically simpler and cheaper practices. In Matiguás–Río Blanco, for example, extremely poor households established 60 ha of fodder banks (38 per cent of the total), and poor households another 37 ha (23 per cent).[11] Extremely poor households also established 153 ha (40 per cent) of pastures with high tree density, with poor households providing another 78 ha (20 per cent). Similarly, in Quindío, low income households adopted many complex land uses, including 70 ha of pastures with high

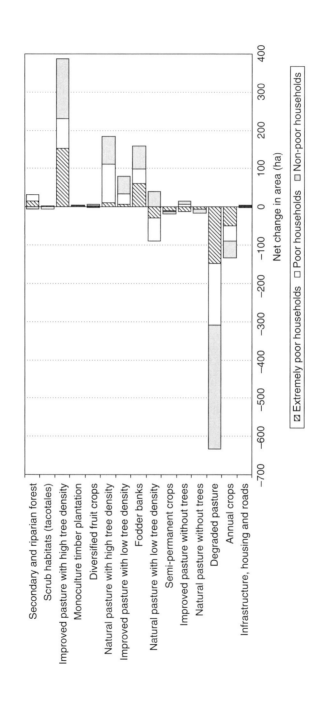

Source: Authors' computations from Silvopastoral Project mapping data.

Figure 9.1 Land use change during Silvopastoral Project, by income group, Matiguás–Río Blanco, Nicaragua

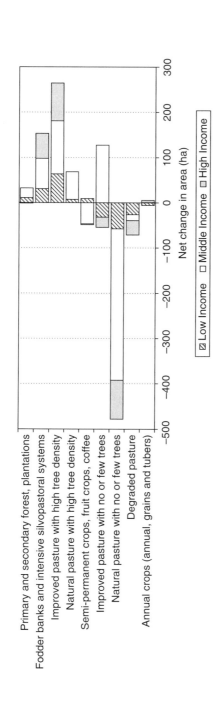

Source: Authors' computations from Silvopastoral Project mapping data.

Figure 9.2 Land use change during Silvopastoral Project, by income group, Quindio, Colombia

tree density and 31 ha of fodder banks and intensive silvopastoral systems (*Leucaena* planted at 5000 trees/ha). Indeed, it was the middle income group that made the simplest possible changes on almost half of the land they converted, replacing natural with improved grasses (mainly star grass, *Cynodon plectostachyus*) in pastures with few or no trees.

Tables 9.5 and 9.6 examine various indices of household participation across income groups. In terms of area converted, poorer households perform poorly at both sites. At Matiguás–Río Blanco, non-poor households converted just under 20 ha each, on average, almost double the 10 ha converted by extremely poor households; while in Quindío, high income households converted about 34 ha each, over three times the 9 ha converted by low income households. It is interesting to note, however, that poor (but not extremely poor) households in Matiguás–Río Blanco converted practically as large an area as the non-poor (18 ha compared with 20 ha).

Data on total land-use changes by households in each income group are affected by the different land endowments of each group, making the total area converted a poor measure of relative participation. The differences across income groups shrink considerably when changes are expressed in terms of proportion of farm area converted. In Matiguás–Río Blanco, poor households converted the greatest proportion of their farms: 57 per cent, compared with 49 per cent for non-poor households. Although extremely poor households converted the smallest proportion of their farms (42 per cent) of any of the groups, the difference was less marked than in absolute terms. In Quindío, high income households converted the greatest proportion of their farms (55 per cent) but low and middle income farms were not far behind (40 per cent converted). Moreover, these differences in the proportion of farms converted are not significant (at the 5 per cent level) at either site.

Whether expressed in hectares or in proportion of farm area converted, area-based indicators fail to measure whether the changes are large or small. Sowing improved pasture grasses in a treeless pasture requires substantially less effort (and generates fewer benefits) than converting it to pasture with high tree density, yet will have the same value in terms of either the area converted or the percentage of farm area converted. Area-based indicators also omit investments in live fencing. One option to incorporate a measure of intensity is to weight the area converted by the ESI of the land-use change, and then add points for live fencing. While the ESI is not intended as a measure of effort, higher ESI land uses tend to involve more effort than lower ESI uses. This measure is also appealing as it is the outcome that is of interest to the buyer of the environmental services. The increase in total ESI is the simplest measure (and is readily available, as it

Table 9.5   Participation rates among PES recipients by income group, Matiguás-Río Blanco, Nicaragua

Income group	Total land (ha)	Change in land use		Live fencing		Environmental services index				
						(Total points)		(Points/ha)		Change (%)
		(ha)	(%)	Initial (km)	Increase (km)	Initial	Increase	Initial	Increase	
Per household:										
Extremely poor	23.7	10.0	42.2	0.93	1.96	19.0	7.9	0.80	0.33	41.6
Poor	30.9	17.7	57.4	1.24	2.32	21.4	15.3	0.69	0.50	71.4
Non-poor	40.1	19.5	48.8	1.77	2.73	28.4	16.0	0.71	0.40	56.1
All	30.3	14.6	48.2	1.26	2.28	22.5	12.0	0.74	0.39	53.2
Total area:										
Extremely poor	1041.9	440.1	42.2	40.81	86.44	837.0	348.5	0.80	0.33	41.6
Poor	586.5	336.6	57.4	23.53	44.02	407.3	290.7	0.69	0.50	71.4
Non-poor	1161.6	566.8	48.8	51.46	79.20	824.9	462.6	0.71	0.40	56.1
All	2790.0	1343.5	48.2	115.80	209.66	2069.3	1101.8	0.74	0.39	53.2

Notes: Totals may not add up because of rounding.

Source: Computed from Silvopastoral Project mapping data.

Table 9.6 Participation rates among PES recipients by income group, Quindio, Colombia

Income group	Total land (ha)	Change in land use		Live fencing		Environmental services index				
		(ha)	(%)	Initial (km)	Increase (km)	(total points) Initial	Increase	(points/ha) Initial	Increase	Change (%)
Per household:										
Low income	23.3	9.4	40.4	0.02	3.62	14.6	8.1	0.63	0.35	55.3
Middle income	49.2	19.8	40.3	0.04	5.38	34.2	13.9	0.69	0.28	40.7
High income	62.1	34.2	55.0	0.04	6.55	38.4	25.9	0.62	0.42	67.3
All	40.2	17.5	43.5	0.03	4.81	26.6	13.1	0.66	0.33	49.4
Total area:										
Low income	698.8	282.5	40.4	0.49	108.71	437.4	241.7	0.63	0.35	55.3
Middle income	1573.3	633.8	40.3	1.21	172.03	1093.1	444.8	0.69	0.28	40.7
High income	621.1	341.6	55.0	0.37	65.54	384.5	258.7	0.62	0.42	67.3
All	2893.2	1257.9	43.5	2.07	346.28	1915.0	945.2	0.66	0.33	49.4

Notes: Totals may not add up because of rounding.

Source: Computed from Silvopastoral Project mapping data.

forms the basis for payments to participants), but, like the area converted, it is constrained by total farm size. Examining the increase in ESI per hectare or the percentage increase in ESI addresses this problem.

In Matiguás–Río Blanco, the increase in ESI points is greatest in absolute terms for non-poor households, but poor households follow close behind, and the difference is not statistically significant. Poor households do even better in proportional change in ESI. Extremely poor households trail, but the difference is not statistically significant. Extremely poor households had the highest initial ESI/ha, so they may have had less scope for substantial improvements. In Quindío, the absolute increase in ESI points is smallest for low income households and largest for high income households, but, in relative terms, the 55 per cent increase achieved by low income households exceeded the 41 per cent achieved by middle income households and rivalled the 67 per cent of high income households. Once again, these differences are not statistically significant.

Examination of observed land-use changes thus indicates that poorer households are, in fact, able to participate quite extensively in the Silvopastoral Project's PES scheme, even though it requires some technically complex and onerous land-use changes. Participation rates by poorer households are broadly similar to those of better-off households at both sites: lower by some measures, but higher by others. To shed further light on participation decisions and the factors that may affect them, we undertook econometric analyses of participation rates at both sites.

The literature on adoption decisions usually looks at the binary choice of whether or not to adopt a given practice, using cross-sectional data on adopters and non-adopters, and the effect of different factors on the probability of adoption (Pattanayak et al. 2003). This approach is not relevant in our case because project funding limited the number of participants and because a binary adoption/non-adoption choice would fail to capture the nature of participation in the project. Rather than participation per se, what is of interest here is how household characteristics affect the intensity of participation, with a particular focus on whether poorer households are less able to participate than better-off households. Our approach is similar to that of Nkonya et al. (1997), who examined the intensity of adoption of improved seed in Tanzania using continuous variables (hectares planted with improved maize seed or amount of fertilizer applied per hectare of maize), and of Rajasekharan and Veeraputhran (2002), who examine the share of farms using intercropping in Kerala, India.

We ran five different regressions for each site, using the indices of participation discussed earlier (area converted, share of farm converted, absolute change in ESI points, percentage change in ESI points and change in ESI points/ha). Like Rajasekharan and Veeraputhran (2002),[12] we employ a one-tailed Tobit to model farm area, as this variable is restricted to non-negative values. We employ a two-tailed Tobit model to model the percentage of the farm area converted, as this ranges between 0 and 100. Changes in ESI, in ESI per hectare and percentage changes in ESI can take any value and so are modelled using ordinary least squares (OLS).

Our choice of explanatory variables draws on the factors identified by Pagiola et al. (2005) as likely to affect participation in PES, and by Pattanayak et al. (2003) and the studies they cite as likely to affect adoption of agroforestry practices, as discussed below. The number of explanatory variables that could be included was limited by the relatively small number of observations at both sites. In this case, increasing the number of observations was not an option: our data include every single PES recipient at both sites. Fortunately, the small size of the sites means that many potential explanatory variables vary little across households and thus can be safely omitted.

Tables 9.7 and 9.8 present the estimation results for the two sites.[13] The first two columns in each table report the results of Tobit models for the area changed and the proportion of farm changed, and the last three columns show the results for the OLS models for the change in ESI, the percentage change in ESI and the change in ESI per hectare. Measures of model fit are relatively low, but this is not surprising with cross-sectional data, particularly when sample sizes are small. They are comparable with those obtained by Rajasekharan and Veeraputhran (2002) and Ervin and Ervin (1982).

The first group of independent variables examines the effect of farm characteristics. Many previous studies report a positive effect of farm size on adoption of various practices, which has been interpreted as indicating higher flexibility of the farming system or the existence of economies of scale (Nowak 1987; Thacher et al. 1997; Rajasekharan and Veeraputhran 2002). In these results, farm area is positively associated with intensity of adoption measured in area converted at both sites, but has a small and non-significant impact on most other indicators of participation. This suggests that the correlation between farm size and area converted is simply due to larger farms having more area to convert.

Labour availability would seem likely to be important, although it is seldom significant (Pattanayak et al. 2003). A measure of the hours per week worked on the farm was included. Family labour is significant and positive in the farm area model in Quindío, but non-significant in the farm share and ESI models, and in all the models in Matiguás–Río Blanco. This is not surprising, as the relationship between land-use change and labour use is complex: switching to higher ESI land uses does not necessarily increase labour use. Interestingly, the age of the household head has

Table 9.7   Estimation results, Matiguás-Río Blanco, Nicaragua

Independent variable	Dependent variable				
	Area changed (ha)	Proportion of farm changed (%)	Change in ESI (points)	Change in ESI (%)	Change in ESI per ha
Model:	Tobit	Tobit	OLS	OLS	OLS
Constant	−1.085	41.370***	54.572	0.137	−8.310**
	(3.997)	(9.622)	(35.037)	(0.110)	(3.883)
Farm area (ha)	0.374***	0.068	0.176	0.002	0.410***
	(0.075)	(0.094)	(0.318)	(0.001)	(0.081)
Family labour	−0.039	−0.219**	−0.612	−0.003**	−0.105
(hours/week/ha)	(0.072)	(0.109)	(0.380)	(0.001)	(0.075)
Livestock units	−0.080	0.314	−0.355	−0.000	0.076
	(0.165)	(0.566)	(1.456)	(0.006)	(0.170)
Experience (years)	−0.203**	−0.583***	−1.779***	−0.005**	−0.169**
	(0.084)	(0.152)	(0.620)	(0.002)	(0.082)
Male-headed	1.792	−1.548	14.579	0.079	3.283
household (1 = yes)	(2.750)	(4.725)	(18.895)	(0.048)	(2.675)
Year-round access	−0.783	3.860	36.796**	0.145**	3.263*
by road (1 = yes)	(2.703)	(5.853)	(17.804)	(0.061)	(1.784)
Hilly topography	−0.004	0.073	−0.021	0.000	−0.035
(% farm area)	(0.023)	(0.053)	(0.204)	(0.001)	(0.026)
Access to credit	2.127	5.998*	12.713	0.079*	2.304
(1 = yes)	(1.546)	(3.261)	(14.034)	(0.043)	(1.481)
Income share of	0.081**	0.235**	1.970***	0.012***	0.188***
off-farm job	(0.040)	(0.103)	(0.389)	(0.001)	(0.041)
Technical	−1.114	−2.147	−6.648	−0.058	−1.616
assistance from	(1.650)	(3.652)	(13.664)	(0.043)	(1.541)
project (1=yes)					
PES recipient	7.579***	11.744***	−7.395	0.103*	6.548**
(1 = yes)	(2.432)	(4.548)	(17.829)	(0.059)	(2.534)
Poor (1=poor)	1.753	1.797	18.599	0.075	3.465
	(2.392)	(5.287)	(22.324)	(0.066)	(2.421)
Extremely poor	−2.700	−5.444	−0.962	0.004	−1.761
(1 = extremely poor)	(1.812)	(4.252)	(15.576)	(0.057)	(1.941)
R^2			0.016	0.092	0.527
Pseudo R^2	0.639	0.211			
Number of observations	116	116	116	116	116

Notes: Standard errors in parentheses; robust standard errors for OLS coefficients.
*, **, *** indicates coefficient estimate is significantly different from zero at 90 per cent,
95 per cent, or 99 per cent confidence level.

Payments for environmental services

Table 9.8   Estimation results, Quindío, Colombia

Independent variable		Dependent variable				
		Area changed (ha)	Proportion of farm changed (%)	Change in ESI (points)	Change in ESI (%)	Change in ESI per ha
	Model:	Tobit	Tobit	OLS	OLS	OLS
Constant		−20.433*	26.779	−12.931**	45.900	0.255*
		(10.755)	(20.745)	(6.375)	(29.207)	(0.141)
Farm area (ha)		0.423***	−0.005	0.181***	−0.366***	−0.002***
		(0.070)	(0.054)	(0.036)	(0.097)	(0.000)
Livestock units		0.052	0.004	0.087***	0.306***	0.002***
		(0.036)	(0.044)	(0.020)	(0.086)	(0.000)
Family labour (adults/ha)		9.844**	21.615	3.886	10.923	0.065
		(4.451)	(14.450)	(2.747)	(19.461)	(0.103)
Experience (years)		0.119	0.024	0.098	−0.210	−0.002
		(0.089)	(0.254)	(0.062)	(0.341)	(0.002)
Male-headed household (1 = yes,)		6.202	17.067	1.858	−21.439	−0.065
		(3.837)	(10.597)	(2.306)	(20.936)	(0.074)
Distance to nearest village (km)		−0.447*	−1.407***	−0.205	−0.428	−0.006
		(0.242)	(0.517)	(0.179)	(0.946)	(0.005)
Flat topography (% farm area)		−0.028	−0.083	−0.016	−0.263**	−0.001**
		(0.043)	(0.088)	(0.030)	(0.126)	(0.001)
Assets (1000 COP)		−0.000*	−0.001**	−0.000	−0.001*	−0.000*
		(0.000)	(0.000)	(0.000)	(0.000)	(0.000)
Income share of off-farm job		3.602	−6.877	−0.286	−22.742	−0.136
		(8.430)	(29.262)	(5.106)	(42.335)	(0.287)
Technical assistance from project (1 = yes)		2.850	9.350	3.630**	15.007	0.079
		(1.961)	(5.995)	(1.504)	(12.790)	(0.065)
PES recipient (1 = yes)		7.957**	18.360**	6.355***	42.998***	0.294***
		(3.110)	(8.259)	(1.547)	(11.031)	(0.059)
Low income (1 = low income)		0.774	−11.424	0.905	2.824	−0.020
		(6.144)	(8.509)	(4.315)	(13.441)	(0.068)
Middle income (1 = middle income)		0.054	−14.135*	0.680	3.445	0.008
		(5.782)	(7.726)	(3.843)	(11.200)	(0.060)
R^2				0.794	0.298	0.382
Pseudo R^2		0.83	0.22			
Number of observations		101	101	101	101	101

Notes: Robust standard errors in parentheses.
*, **,*** indicates coefficient estimate is significantly different from zero at 90 per cent, 95 per cent, or 99 per cent confidence level.

a consistent, statistically significant negative impact on intensity of participation in all models in Matiguás–Río Blanco, but generally small and non-significant impacts in Quindío. Other studies have often found a positive effect, though rarely a significant one, and have generally attributed it to experience reducing the risks of adoption (Pattanayak et al. 2003); an alternative explanation, and one consistent with the Matiguás–Río Blanco results, is that older farmers may be less inclined to make changes. Whether households were male-headed does not have a significant impact on participation at either site under any formulations.

The second group of independent variables concerns factors likely to affect the profitability of adoption. As these study areas are small (particularly Matiguás–Río Blanco), most farms face similar prices for inputs and outputs and have similar yield potentials. The profitability of various silvopastoral practices should therefore be broadly similar throughout each area. Farmers with lower accessibility will tend to face higher input costs and lower output prices at the farm gate. Indeed, distance from the nearest village has a significant negative impact on the extent of the area converted in Quindío, though the impact on ESI is not significant. In Matiguás–Río Blanco, where distances are smaller but the roads are worse, whether farms have year-round access had strong positive impacts on changes expressed in ESI, but not on area-based indicators.

The proportion of the farm on flat terrain has a negative impact in Quindío, but it is not significant except in some of the ESI models. In Matiguás–Río Blanco this variable was expressed as a proportion of the farm on hilly terrain, but was found not to be significant in any model. In general, there is no strong a priori reason to expect a particular sign on topography variables. Land on steep slopes may benefit more from silvopastoral practices because it is more vulnerable to degradation under traditional extensive grazing, but the cost of implementing practices may be higher.

In Quindío, herd size has a significant positive impact in the ESI models but not in the area models, suggesting that its impact is primarily through its demand for fodder rather than through its contribution to financing. Herd size is not significant in any of the Matiguás-Río Blanco models, perhaps because lower stocking rates make demand for fodder less of a constraint.

The third group of independent variables includes factors that have been hypothesized to affect the ability of households to participate in the scheme.[14] The ability to finance the necessary investments is one potential obstacle. To examine whether initial investment costs affect the ability to participate, a measure of assets was included in the Quindío analysis. In Matiguás–Río Blanco, where initial assets vary relatively little across households, access to credit was used, measured as a binary indicator of whether a household had access to credit during the 5 years before project implementation.[15] Off-farm income was measured as the income share of off-farm jobs held by all household members. Off-farm income can be a financing source for investment in new practices, but can also result in a higher opportunity cost of labour.

In Matiguás–Río Blanco, both access to credit and off-farm income had a positive impact on intensity of participation in every model, although only that of off-farm income was significant. The non-significance of credit is surprising, given the cost of implementing some of the practices promoted by the Silvopastoral Project and the low income levels of most households. The first-year survey of participants provided the explanation for this result: even in poor areas such as Matiguás–Río Blanco, most households had a variety of ways to finance investments (Pagiola et al. 2008). Some investments were undertaken entirely with family labour and so did not require financing. Unsurprisingly in a livestock-producing area, the sale of animals was the most frequently mentioned source of funds (61 per cent of all households). The initial 'baseline' payment also played an important role for many households (53 per cent). Although almost no households mentioned off-farm income as a source of financing, it is possible that such income contributed primarily through its contribution to savings, which was an important source of financing for 41 per cent of households.

In Quindío, household assets had a very small but negative sign, in all formulations, but they were not significant. Off-farm income had a mostly negative, but not significant, effect on adoption intensity in Quindío. These results are consistent with an interpretation of off-farm employment increasing the opportunity cost of labour. Savings were the most often cited source of financing for first-year investments in Quindío, followed by animal sales; baseline payments played a very small role, probably because low initial ESI points meant that these payments were small.

The technical difficulty of adopting silvopastoral practices is the other main potential obstacle to household participation. To test the importance of this factor, the project provided technical assistance to a randomly selected subset of PES recipients. The results were strikingly different across the two sites. In Matiguás–Río Blanco, access to the Project's TA was not significant under any formulation, while in Quindío it was significant under every formulation. The explanation for the lack of significance in Matiguás–Río Blanco is probably two-fold: first, as noted above, silvopastoral practices were already relatively well known in the area before the project. Second, the very small size of the site probably made it much easier for households who did not receive TA directly to learn from their neighbours who did. In contrast, silvopastoral practices were practically unknown in Quindío before the project, and the larger size of the site (and, hence, the much lower density of recipients in the overall farm population) made TA play a much more important role.

The strongest result in the Quindío models was that being a PES recipient had a large positive impact on the extent of adoption of silvopastoral measures, irrespective of how adoption was measured. This confirms the observed sharp contrast in land-use change between PES recipients and control group measures, and indicates that the contrast was not due to self-selection of strongly motivated land users into the scheme or to differences in characteristics across groups. The Matiguás–Río Blanco results also show a strong positive impact of being a PES recipient, but given the concerns over the quality of the control group, this result cannot be treated as more than suggestive.

As poverty is multidimensional, dummies for the income groups were also included, with the highest income group omitted, to capture other aspects of poverty that may not be incorporated in the previous variables. These dummies show some interesting patterns at both sites. In Matiguás–Río Blanco, poor (but not extremely poor) households had higher levels of participation than non-poor households in every model. Extremely poor households, on the other hand, had lower levels of participation than non-poor households in terms of the area converted and in most ESI models. None of these differences was statistically significant, however. These results confirm the patterns seen in Table 9.5. In Quindío, both the low income and the middle income group dummies have a mix of positive and negative coefficients, but, except in one case, the effect was not statistically significant. This suggests that income level has relatively little impact on participation that is not already captured by other variables (such as farm size) which may be correlated with poverty. The only significant impact is that middle income households convert a lower share of their farm than high income households. This may be due to that group using a portion of their farm for other productive activities; indeed, low income households actually expanded the area under shade-grown coffee to a small extent, and replaced part of their monoculture fruit crop areas with diversified fruit crops.

CONCLUSIONS

Can poorer households participate in PES scheme? The experience of the Silvopastoral Project in both Matiguás–Río Blanco and Quindío indicates that they can. Not only did poorer households participate quite extensively, but by some measures they participated to a greater extent than better-off households in Matiguás–Río Blanco. Participation by poorer households was somewhat lower in Quindío, but the differences were not statistically significant. Even the poorest households at both sites participated at high rates in the project, and their participation was not limited solely to the simpler and cheaper practices. These results are particularly strong because the Silvopastoral Project imposes much greater burdens on participants than most PES schemes. They bode well for the prospects of poor households being able to participate in PES schemes financed from Reduced Emissions for Deforestation and forest Degradation (REDD) payments, as these schemes are likely to focus on much simpler and less costly forest conservation measures.

This conclusion obviously needs to be approached with some caution. It is possible that the high levels of participation by poorer households are due to self-selection bias, where only those households able to participate joined. We believe that this is unlikely for two reasons. First, the project offered a very wide range of participation options, including many that are not very onerous, even for poorer households. Indeed, households could, in principle, have done absolutely nothing; they would then have received the baseline payment but would not have received any payment beyond that. In fact, no household chose that route. Second, many non-participating households at both sites wanted to participate as well, but were prevented from doing so by the project's own limits on the number it could accept. Even if there were some self-selection at play, it is significant that in a poor area such as Matiguás–Río Blanco there were many poor households – including many extremely poor households – that were able to participate in the scheme, and even to undertake expensive and technically challenging land-use practices.

Nevertheless, one should not jump to the sanguine conclusion that all poor farm households everywhere will always be able to participate in PES schemes.[16] Both PES schemes and local conditions differ from case to case, and there may well be cases where otherwise eligible poor households may find it difficult or impossible to participate. Indeed, results at both sites show that the poorest households – although by no means shut out – do appear to have had greater difficulty in participating as intensively as less poor households.

These detailed results help to identify several specific factors that tend to affect participation. This information can help designers of PES schemes to reduce potential obstacles to the participation of poorer households. There is little that can be done about poorer farms being less accessible, but designers can do something about financing constraints and technical difficulty.

The significance of credit underlines an important potential constraint for poorer households. This constraint will not always be present in PES schemes. When schemes require maintaining existing practices – as in the majority of contracts in Costa Rica's scheme, for example, and as in prospective REDD-financed schemes – there are few or no investment requirements.[17] Financing constraints may be important when land-use changes are required for participation. However, these results suggest that this constraint is not absolute, as it is sometimes made out to be. Even poor households, such as those in Matiguás–Río Blanco, often have a variety of ways to finance profitable investments. Nevertheless, it is likely that poorer households will have fewer alternatives: fewer savings, fewer assets that might be sold and/or worse access to credit. Providing some initial financing (such as the baseline payment made by the Silvopastoral Project) may be desirable for PES schemes that involve initial investments in areas with many poor households.[18]

Our results also highlight the need to understand whether TA may be required. The need for TA appears to be linked more with previous experience than with poverty. When PES schemes require participants to adopt relatively well-known practices, TA may be of minor importance. Conversely, when participation requires adoption of practices that are not widely used, even relatively well-off households such as those in Quindío may need TA.

The availability of multiple options in the Silvopastoral Project may well have contributed to high participation by poorer households, as they were able to choose the options that work best for them, in light of their particular constraints. When there are multiple ways of providing a service (or different levels of a service), it makes sense to offer multiple ways in which households can participate, as long as transaction costs do not increase unduly. It is interesting to note, however, that at these sites, the poorest households did not choose the cheaper and easier land uses: in fact, it was the better-off households that did so in Matiguás–Río Blanco, and the middle income group that did so in Quindío.

In general, transaction costs are likely to be a bigger threat to the participation of poorer households in PES schemes than their own ability to participate (Pagiola et al. 2005). The results presented above illustrate this. As can be seen in Table 9.6, low income households in Quindío converted 40 per cent of their farms, on average, and increased their ESI score by about 55 per cent. These participation rates were not far below those of high income households, who converted 55 per cent of their farms and increased their ESI score by 67 per cent. From the perspective of the service buyer, what matters is the total absolute increase in environmental service generation (whether proxied by area, as is commonly the case, or by more sophisticated measures such as the ESI), and the unit cost of achieving it. The cost, in turn, has two components: the cost of the payment, which is identical for a given increase in ESI for all households, and the transaction cost of contracting with each household. This second cost is likely to be largely fixed per household, irrespective of farm size. High income households converted a total of 342 ha and achieved a total increase in ESI of 259 points. At first glance, the results for low income households appear similar, as they converted a total of 283 ha and achieved a total increase in ESI of 242 points. The number of low income households is much larger, however, so that on a per contract basis, the comparison is less favourable: low income households converted 9.4 ha and increased ESI by 8.1 points per contract, while high income households converted 34.2 ha and increased ESI by 25.9 points per contract. In other words, it takes more than three contracts with low income households to achieve the same results as a single contract with a high income household. Very similar relationships can be seen in the Matiguás–Río Blanco (Table 9.5). The larger the transaction costs, the more attractive it will be for PES schemes to focus on larger land holdings, and as farm size tends to be highly correlated with income, in practice this will mean focusing on less poor households. This is not a purely hypothetical concern: in Ecuador, the PROFAFOR scheme has decided to adopt a 50 ha minimum size for the forest plantations from which it buys carbon sequestration services (Wunder and Albán 2008).

Keeping transaction costs low – in addition to being desirable in itself – is thus imperative if poorer households are not be shut out of many PES schemes.[19] However, the smaller farm size of many poor households means they will always have relatively higher transactions costs. It is thus important to attempt to devise mechanisms to overcome them. Costa Rica experimented with collective contracting, under which groups of small farmers joined the country's PES scheme collectively, rather than individually, spreading transaction costs over a large group. This approach ran into problems, however, as non-compliance by a single group member resulted in payments being halted to all members. The approach has thus been revised to process the applications of such groups together, but then issue individual contracts; this avoids the partial compliance problem, but has much smaller savings in transaction costs (Pagiola 2008). This is clearly an area in which more work, and some imaginative solutions, will be necessary. This is also an area in which development aid could be used to leverage PES schemes by providing support to the participation of poorer households, and, in particular, by underwriting some of the transaction costs involved.

ACKNOWLEDGMENTS

Work on this paper was financed in part by a grant from the Norwegian Trust Fund for Environmentally and Socially Sustainable Development. The Nicaragua participant survey was conducted by Alfredo Ruiz García, María Raquel López Vanegas and Yuri Marín López, and the Colombia participant survey was conducted by a team from CIPAV led by Daniel Uribe. We have benefited from discussions with Elías Ramírez, Enrique Murgueitio R., Alvaro Zapata Cadavid, Andrés Felipe Zuluaga, Paola Agostini, José Gobbi, Cees de Haan, Muhammad Ibrahim, Mauricio Rosales and Juan Pablo Ruíz. Any remaining errors are the authors' sole responsibility. The opinions expressed in this paper are the authors' own and do not necessarily reflect those of the World Bank Group or the Inter-American Development Bank.

NOTES

1. Wunder (2008) adds a fourth 'filter': whether households are competitive in terms of transaction costs. This filter affects whether the PES scheme will select particular households and is thus closely related to eligibility criteria. We return to this issue in the conclusion.
2. In this context, the GEF can be considered to be buying services on behalf of the global community.
3. The ESI is described in detail in CIPAV (2003) and Pagiola et al. (2004).
4. Payments of US$75/ESI point imply a payment for carbon of US$7.50/ton Carbon (tC), which compares favourably to world prices at the time of project launch of US$14–20/tC (World Bank 2004). A similar comparison for biodiversity is difficult. The highest possible payment was US$300/ha over 4 years, equivalent to about US$30/ha/year on an annualized basis, which is less than similar payments under Costa Rica's and Mexico's PES schemes.
5. The Silvopastoral Project's use of short-term payments is controversial, as payments in PES schemes should generally be ongoing. The use of short-term payments means that the conditionality of the scheme is limited to the first few years, which may well affect its long-term viability (Pagiola et al. 2007). This debate does not affect the issue considered in this chapter, which focuses on the extent of participation and whether poverty affects it.
6. It is important to note that this initial 'baseline' payment was intended as recognition of the environmental services that households were already providing, and not as subsidy to increased service provision. Households receiving the baseline payment were under no obligation to participate further. As noted below, however, the baseline payment did help finance some of the required investments, particularly in Matiguás–Río Blanco. It also played a very important role in establishing trust – that is, that participants would indeed receive the promised payments if they undertook the land-use changes the project was asking for.
7. Livestock are converted into livestock units (*Unidad Gran Ganado*, UGG) using the following conversion factors: adult cows, 1.0 UGG; oxen or breeding bulls, 1.55 UGG; calves, 0.33 UGG; yearlings, 0.7 UGG.
8. At the beginning of the project, US$1 = C$14.25 and US$1 = COP2670.

9. Household income was computed by adding all income sources reported by participants, including net income from agricultural, forest and dairy production; livestock sales; off-farm work; net income from non-farm enterprises; and remittances. Dairy, agricultural and forest products consumed by the household were included in the calculation of income using market prices, and the value of family labour was imputed using local wage rates for unskilled labour. Expenditure is generally preferred over income as an indicator of household welfare, as it tends to be less variable (Ravallion 1992). However, the baseline survey only collected data allowing income to be computed. Moreover, these data are based largely on information self-reported by the farmers, and so are subject to both recall problems and possible biases. These biases are unlikely to affect results as long they are similar across income groups.

10. The figures quoted show net changes, and thus understate the extent of change. In addition, existing silvopastoral practices were often upgraded to more intensive practices (for example, increasing the density of trees in pastures).

11. The popularity of fodder banks among poorer households in Matiguás–Río Blanco may be due to the greater availability and lower opportunity cost of labour in such households. Substantial amounts of labour are required by the cut-and-carry practices that fodder banks imply.

12. Rajasekharan and Veeraputhran (2002) employed a one-tailed Tobit to study the adoption of intercropping in three regions in Kerala, India, using the share of farm area under intercropping as the dependent variable.

13. OLS models were tested for heteroskedasticity in the error distribution using the Breusch-Pagan test. Results for these tests rejected the null hypothesis of homoskedasticity of errors, which, if ignored, would result in the loss of optimality of the OLS estimator (Greene 2000; Mittelhammer et al. 2000). In the absence of prior information about the structure of the heteroskedasticity, we used the OLS estimator with White's heteroskedasticity-consistent covariance matrix estimator (White 1980). We found no evidence of either moderate or strong multicollinearity in any of the regression models using the Belsley et al. (1980) diagnostics in OLS models and the Belsley (1991) diagnostic in the Tobit models.

14. Within these factors, tenure is not an issue at either site, as both were selected partly for the absence of such problems.

15. The endogeneity of credit was tested using the Wu-Hausman and Hausman tests in the OLS models and the Smith-Blundell test in the Tobit models. Exogeneity of credit was not rejected in any model at the 90 per cent confidence level.

16. It is important to recall that this case study does not speak to possible differences in eligibility to participate, due to spatial considerations or tenure problems. Pagiola and Colom (2006) find that the areas in Guatemala that are important for the provision of water services do not always have high poverty rates.

17. Forest conservation may require out-of-pocket expenditures by participating households, for example to fence off forests receiving conservation contracts or undertake measures to prevent forest fires. However, these costs are much lower than those associated with land-use changes.

18. Costa Rica front-loads payments under its timber plantation contract for this reason. Front-loaded payments, however, introduce other problems, as they reduce the conditionality of the scheme.

19. The Silvopastoral Project as it was conducted is a poor example of this, as it had relatively high monitoring costs dictated in part by its pilot nature and in part by the need to distinguish small differences in land use so as to compute ESI scores.

REFERENCES

Belsley, D.A. (1991), *Conditioning Diagnostics, Collinearity and Weak Data in Regression*, New York: John Wiley & Sons.
Belsley, D.A., E. Kuh and R.E. Welsch (1980), *Regression Diagnostics: Identifying Influential Data and Sources of Collinearity*, New York: John Wiley & Sons.
Bruijnzeel, L.A. (2004), Hydrological functions of moist tropical forests: not seeing the soil for the trees? *Agriculture, Ecosystems and Environments* **104** (1), 185–228.
CIPAV (2003), *Usos de la Tierra en Fincas Ganaderas: Guía para el Pago de Servicios Ambientales en el Proyecto Enfoques Silvopastoriles Integrados para el Manejo de Ecosistemas*, Cali: Fundación CIPAV.
Dagang, A.B.K. and P.K.R. Nair (2003), Silvopastoral research and adoption in Central America: Recent findings and recommendations for future directions, *Agroforestry Systems* **59** (2), 149–55.
de Janvry, A. and E. Sadoulet (2000), Rural poverty in Latin America: determinants and exit paths, *Food Policy* **25** (4), 389–409.
Dennis, P., L. Shellard and R. Agnew (1996), Shifts in arthropod species assemblages in relation to silvopastoral establishment in upland pastures, *Agroforestry Forum* **7** (3), 14–21.
Engel, S., S. Pagiola and S. Wunder (2008), Designing payments for environmental services in theory and practice: an overview of the issues, *Ecological Economics* **65** (4), 663–74.
Ervin, C.A. and D.E. Ervin (1982), Factors affecting the use of soil conservation practices: hypothesis, evidence, and policy implications, *Land Economics* **58** (3), 278–92.
Feder, G., R.E. Just and D. Zilberman (1985), Adoption of agricultural innovations in developing countries: a survey, *Economic Development and Cultural Change* **33** (2), 255–95.
Ferraro, P.J. and S.K. Pattanayak (2006), 'Money for nothing? A call for empirical evaluation of biodiversity conservation investments, *PLoS Biology* **4** (4), 482–8.
Fisher, M.J., I.M. Rao, M.A. Ayarza, C.E. Lascano, J.I. Sanz, R.J. Thomas and R.R. Vera (1994), Carbon storage by introduced deep-rooted grasses in the South American savannas, *Nature* **371** 6494, 236–8.
Greene, W.H. (2000), *Econometric Analysis*, 4th edition, Upper Saddle River: Prentice Hall.
Harvey, C. and W. Haber (1999), Remnant trees and the conservation of biodiversity in Costa Rican pastures, *Agroforestry Systems* **44** (1), 37–68.
López, R. and A. Valdés (2000), *Rural poverty in Latin America*, New York: St. Martin's Press.
Mercer, D.E. (2004), Adoption of agroforestry innovations in the tropics: A review, *Agroforestry Systems* **61** (1), 311–28.
Mittelhammer, R.C., G.G. Judge and D.J. Miller (2000), *Econometric Foundations*, New York: Cambridge University Press.
Nkonya, E., T. Schroeder and D. Norman (1997), Factors affecting adoption of improved maize seed and fertiliser in Northern Tanzania, *Journal of Agricultural Economics* **48** (1–3), 1–12.
Nowak, P.J. (1987), The adoption of agricultural conservation technologies: economic and diffusion explanations, *Rural Sociology* **52** (2), 208–20.

Pagiola, S. (2008), Payments for environmental services in Costa Rica, *Ecological Economics* **65** (4), 712–24.

Pagiola, S. P. Agostini, J. Gobbi, C. de Haan, M. Ibrahim, E. Murgueitio, E. Ramírez, M. Rosales and J.P. Ruíz (2004), Paying for biodiversity conservation services in agricultural landscapes, Environment Department Paper No. 96, Washington DC: World Bank.

Pagiola, S., A. Arcenas and G. Platais (2005), 'Can payments for environmental services help reduce poverty? An exploration of the issues and the evidence to date from Latin America', *World Development*, **33** (2), 237–53.

Pagiola, S. and G. Platais (2007), *Payments for Environmental Services: From Theory to Practice*, Washington DC: World Bank.

Pagiola, S. and A. Colom (2006), Will payments for watershed services help reduce poverty? Evidence from Guatemala, Unpublished paper, Washington DC: World Bank.

Pagiola, S., E. Ramírez, J. Gobbi, C. de Haan, M. Ibrahim, E. Murgueitio and J.P. Ruíz, (2007), Paying for the environmental services of silvopastoral practices in Nicaragua, *Ecological Economics* **64** (2), 374–85.

Pagiola, S., A. Rios and A. Arcenas (2008), Can the poor participate in payments for environmental services? Lessons from the Silvopastoral Project in Nicaragua, *Environment and Development Economics* **13** (3), 299–325.

Pattanayak, S.K., D.E. Mercer, E. Sills and J.-C. Yang (2003), Taking stock of agroforestry adoption studies, *Agroforestry Systems* **57** (3), 173–86.

Rajasekharan, P. and S. Veeraputhran (2002), Adoption of intercropping in rubber smallholdings in Kerala, India: a tobit analysis, *Agroforestry Systems* **56** (1), 1–11.

Ravallion, M. (1992), Poverty comparisons: a guide to concepts and methods, LSMS Working Paper No. LSM88, Washington DC: World Bank.

Swallow, S., M. van Noordwijk, S. Dewi, D. Murdiyarso, D. White, J. Gockowski, G. Hyman, S. Budidarsono, V. Robiglio, V. Meadu, A. Ekadinata, F. Agus, K. Hairiah, P. Mbile, D.J. Sonwa and S. Weise (2007), Opportunities for avoided deforestation with sustainable benefits. An interim report of the Alternatives to Slash and Burn Partnership for the Tropical Forest Margins, Nairobi: ASB partnership for the Tropical Forest Margins.

Thacher, T., D.R. Lee and J.W. Schelhas (1997), Farmer participation in reforestation incentive programmes in Costa Rica, *Agroforestry Systems* **35** (3), 269–89.

White, H. (1980), A heteroskedasticity-consistent covariance matrix estimator and a direct test for heteroskedasticity, *Econometrica* **48** (4), 817–38.

World Bank (2004), *State and Trends of the World Carbon Market 2004*, Washington DC: World Bank.

World Bank (2003), Nicaragua reporte de pobreza: Aumentando el bienestar y reduciendo la vulnerabilidad, Report No. 26128-NI, Washington DC: World Bank.

World Bank (2002), Colombia poverty report, Report No. 24524-CO, Washington DC: World Bank.

Wunder, S. (2008), 'Payments for environmental services and the poor: concepts and preliminary evidence', *Environment and Development Economics* **13** (3), 279–97.

Wunder, S. (2005), Payments for environmental services: some nuts and bolts, CIFOR Occasional Paper No. 42, Bogor: CIFOR.

Wunder, S. and M. Albán (2008), Decentralized payments for environmental services: the cases of Pimampiro and PROFAFOR in Ecuador, *Ecological Economics* **65** (4), 685–98.

Zbinden, S. and D.R. Lee (2005), Institutional arrangements for rural poverty reduction and resource conservation, *World Development* **33** (2), 255–72.

APPENDIX 9.1: DATA SOURCES

To examine participation decisions, three data sets were used. The first was the baseline survey conducted in October–November 2002, during project preparation. This survey included very detailed information on household characteristics. A second survey of participants was conducted in March–May 2004, after the first year of project implementation. This survey collected information on land-use changes that occurred in the intervening period. The questionnaires for the surveys are available from the authors on request.

Information from these two surveys was complemented with detailed land-use data for each farm, derived from maps prepared annually between 2003 and 2007 using remote-sensing imagery. Quickbird imagery with a 61 cm resolution was used, providing very high levels of detail. Land-use maps for each farm derived from these images were then extensively ground-truthed to match each plot to one of the 28 different land uses recognized by the project. These maps provided accurate and consistent measures of area.

Both the baseline and follow-up surveys included a control group, as does the mapping dataset. The main intended purpose of this group was to attempt to distinguish project-induced land-use changes from changes induced by other factors (Ferraro and Pattanayak 2006). Pagiola et al. 2008 found that control group members in Matiguás–Río Blanco differed from PES recipients in many important characteristics (such as income, farm size or herd size), and so it was decided that using the control group would not be useful. The Matiguás–Río Blanco control group was retained here for comparison with the Quindío analysis, but the authors caution against placing much significance in the differences between the control group and PES recipients.

All exchange rates and inflation adjustments are based on the World Bank's *World Development Indicators* database.

10. PES schemes' impacts on livelihoods and implications for REDD activities

Luca Tacconi, Sango Mahanty and Helen Suich

INTRODUCTION

The preceding chapters have discussed in detail the impacts of individual PES schemes on livelihoods and the implications for the design of PES schemes. This chapter aims to draw out the main findings from individual cases to answer the two research questions posed in the first chapter, namely:

- What have been the impacts of PES schemes on livelihoods?
- What are the implications for the design of incentive mechanisms for REDD at the local level?

These research questions are addressed with respect to the critical factors, noted in Chapter 1, that influence the risks and opportunities for livelihoods in PES schemes (Landell Mills and Porras, 2003; Pagiola et al., 2005; Corbera et al., 2007a; Pagiola et al., 2008; and Wunder, 2008):

- the nature and location of the environmental service, for instance, the percentage of poor households tends to be higher in remote areas where forests are often located;
- whether people have the recognized and secure resource rights generally needed to enter into PES agreements;
- whether workable regulatory frameworks exist for a specific environmental service;
- how many PES participants are poor and their ability to participate;
- the size of the payment for the provision of the environmental service;

- finance and credit availability for sellers to cover their up-front costs of participation;
- the skills, education, power and negotiating capacity of environmental service sellers;
- availability of good market information and linkages to communication infrastructure; and
- the existence of mechanisms to reduce transaction costs, for example, collective action institutions that facilitate coordination amongst environmental service sellers.

The chapter first discusses the findings about access to PES schemes and their impacts on financial, physical, social, human and natural capital. We focus on common impacts documented across several cases as well as impacts that were perhaps less widespread but that held significance for REDD schemes dealing with large forest areas and communal or state lands. We then consider the implications for the design of REDD activities at the local level. The implications are discussed not only in the context of the case studies presented in the preceding chapters, but also with reference to very recent literature on REDD, published after the research presented in this volume was designed and the preparation of the case studies had begun.

Before proceeding, it should be noted that the lack of quantitative data is still a constraint upon the analysis of the performance of PES schemes and influences our conclusions. This research was initiated with the view that PES schemes around the world had been working for several years and it would be possible to acquire the necessary quantitative data, for instance, on opportunity and transaction costs as well as the environmental services actually generated. However, this did not prove to be the case. This issue points to the need for further quantification and analysis of the performance of PES schemes.

ACCESS TO PES SCHEMES AND IMPACTS ON LIVELIHOODS

Access to PES Schemes

Access to PES schemes may be important to the social sustainability of schemes, particularly where non-participants feel resentful or experience negative impacts from the scheme, as found in the Indonesian and Brazilian cases. All eight studies reported that *some* poor households were able to access PES schemes. The case studies elaborated upon the participation constraints imposed by weak or unclear property rights, as well as high transaction costs and implementation costs, confirming the findings of previous research on PES and livelihoods (summarized earlier and in Chapter 1). However, they also illustrate approaches to reducing these barriers.

Access to a PES scheme is shaped by the criteria used to select target areas for PES schemes, a point which has come up in previous research (see Chapter 1). Some schemes, such as those in Mexico and Brazil, adopted social criteria, in addition to environmental criteria and competitive bidding processes, in determining where to locate schemes. The Mexican scheme went a step further by developing a preferential payment scale that paid higher returns to disadvantaged participants, which may have been a factor in the high level of engagement by poor households reported in that study.

The case studies illustrate two ways that PES schemes can proceed in the absence of private property rights or in the absence of title papers. Schemes that worked with common property, such as in Mexico, or state property, as in Brazil, allowed broader participation based on partial property rights, such as use rights, or recognized customary claims to land. In Brazil, for instance, some 60 per cent of households participating in *Proambiente* lived in extractive reserves located on state land with a 20–30 year concession granted to a community. These PES schemes involved representatives of communities as signatories to the contract and payments were provided to households and/or community bodies. Another PES scheme working on state property (the Philippines) had local government as a signatory, with PES income distributed in the form of projects that had been agreed through community-level planning processes.

Even where land is held under private ownership (for example, in the Colombian/Nicaraguan and Indonesian cases), the requirement to provide land title papers, a fundamental requirement for most schemes, can be complex and involve significant financial and time costs for landowners. This constraint was addressed in the Ugandan case by enabling tenure verification through local authorities, who could verify ownership through records of previous land purchase agreements and wills.

Transaction costs – the time and money involved in the informed negotiation of PES contracts – are another commonly observed constraint to PES participation (summarized in Chapter 1) and were a significant issue in all cases in this book. Projects generally provided a subsidy to cover these costs. A common approach was to channel support through intermediaries (for example, non-government organizations, NGOs), which provided information and assisted participants to navigate contractual arrangements. Later, however, these same intermediaries needed to fund their support work by retaining a share of PES income, which reduced payments to the environmental service sellers.

Another approach to reducing transaction costs was the use of collective rather than individual contracts with many small-holders. Collective contracts were used in schemes that involved private lands, where farmers were contracted on a group basis (for example, in Indonesia, Brazil and Uganda) and for common property and indigenous reserves on state lands, such as in Mexico and Brazil, respectively. The use of collective contracts raises important issues about how revenue flows should be managed (discussed below in the section on financial capital).

Implementation costs have previously emerged as a third key constraint for poorer households entering into PES agreements, in the absence of credit provision (discussed in Chapter 1). In the cases presented in this book, the most significant implementation costs were up-front investments of labour and money by participants, for instance, to introduce new land management practices (Nicaragua/Colombia) and for tree planting (Indonesia, Mozambique). The costs of monitoring – to verify environmental service provision in order to obtain payments – were significant later in the life of these schemes.

The Nicaraguan/Colombian project addressed the issue of up-front costs by offering a range of land-use change options that could be chosen to suit varying household budgets, labour availability and technical capacity, though the authors note that participation was stronger by poor rather than the poorest households in these countries. The other strategy to support up-front implementation costs was front-loading payments to deliver a large proportion of payments early in the scheme. This approach can, however, pose a risk for the long term sustainability of schemes (discussed in the next section).

Financial Capital

The case studies provide new evidence about how PES schemes were contributing to household income, thus filling a gap in existing research because of the recent growth of environmental service markets (see Chapter 1). Some of the PES schemes involved contracts and payments for individual households (Mozambique, Nicaragua/Colombia, Uganda). Other schemes involved contracts with community organizations or village councils (Brazil, Indonesia, Mexico, Philippines), which then managed these payments. In Brazil, Indonesia and Mexico, a proportion of payments was returned to individual households.

Where PES payments were made to households, they were typically a minor component of total household income. The Ugandan and

Mozambican cases highlight that, although small, this income was nevertheless important, as the lump-sum nature of payments enabled investment in things such as land or home improvements, payment of debts and access to medical services. The Mexican case further highlights that the relative significance of PES income was greatest for poorer households. The findings point to the importance of considering the timing and size of payments, as well as their relative significance to households from different social strata.

Collective contracts raise an important question of how PES income is managed. Collective payments to established community institutions enabled investments in infrastructure (Mexico, Philippines) that would not have occurred if household-level payments were used (see also the discussion about physical capital). PES-related employment – for infrastructure construction and forest patrolling – was the second major use of PES-related income. For example, payments to village governments (*barangays*) were used to pay the salaries of forest guards in the Philippines. However, in general, employment opportunities associated with PES schemes were usually short term and the criteria for the distribution of these opportunities were not clearly identified.

This issue should be further considered by PES designers, as inequities in access to employment or community level infrastructure can deepen social disparities and become a point of conflict. In Nepal, for instance, Community Forest User Group investments in infrastructure (electricity, water supply, irrigation) benefited mainly wealthy landholding households, while poor and very poor households gained few or no benefits (Dev and Adhikari 2007). In contrast, investment in social services such as schooling and health, such as those provided in the Mozambican case, tend to be more widely accessible.

The cases presented in this book highlight the potential for collective payments where there are strong and functioning institutions for collective action (for example, Mexico, Philippines), and where collective payments are effectively invested in infrastructure and other community projects. However, monitoring will be necessary to determine who benefits from community infrastructure and facilities. Furthermore, in these two cases, collective action institutions were rooted in a unique sociopolitical context based on a collective indigenous identity. In a different environment, the question of whether collective payments (1) can be effectively managed and (2) provide sufficient incentive to support effective resource management would need to be assessed. These issues will be particularly significant for REDD which, as discussed below, is likely to be implemented in state-owned or common property forest areas that may not have existing and functioning community-level institutions.

Another area that has not been clearly assessed in existing research is how income from PES schemes compares with the overall transaction and opportunity costs faced by farmers. In the cases presented here, the level of payments was not calculated according to opportunity and transaction costs, but was based on input costs and/or the estimated market value of the environmental service. Payments were also often front-loaded in order to cover initial start-up costs for participants, with cash flow significantly diminished, or ending, before the end of the contract period (for example, Mozambique, Uganda). To some extent, this diminishing cash flow was addressed by strategies such as planting useful species (for example, fruit trees) which could supplement PES income without diminishing the environmental service.

However, if participants found that initial payment levels were not sustained, it could lead to non-compliance or the eventual abandonment of the agreement: evidence of this was already apparent only a few years into the project in Indonesia. This risk was amplified by the fact that PES payments did not measure up to the opportunity costs faced by participants. The front-loading of payments may have initially attracted participants, in a sense creating a honeymoon period. But as the opportunity costs became more striking, participants would default on their agreement. This is consistent with the finding that there was a general lack of understanding among participants about how the scheduling and amount of payments actually operated.

In some cases, for instance, Mozambique, the income generated by PES schemes was reported to have had a stimulus effect on the local economy, creating employment in infrastructure construction and micro-enterprise development, with local businesses growing to absorb the increase in local disposable income. Such flow-on effects could not be attributed to project activities alone and a detailed analysis could not be carried out as part of this research, but would be useful to explore in the future.

Human Capital

The skills, education and negotiating capacity of environmental service sellers has emerged as a key factor in whether small-holders can access and draw significant benefits from such schemes (Chapter 1). These case studies highlight ways in which projects enhanced the capacity of sellers to access PES schemes, effectively implement the requirements of PES agreements and manage their PES income.

Capacity-building was considered a key impact in all cases, covering such areas as environmental awareness, land management, agroforestry and silviculture, governance of local representative bodies, business development and the understanding of PES. The projects supported this through intermediary organizations (NGOs or government agencies, often supported by government or donors).

The Mexican case highlights the importance of capacity-building in enabling broad access to PES schemes. However, the intermediaries' lack of skills and knowledge in relevant areas had diminished their ability to design workable projects and to manage revenue flows, which, in turn, contributed to conflict between community members and intermediaries in some sites. There was little evidence on the long-term impact of capacity-building activities, in particular, on whether new knowledge and skills were applied in practice. Persisting confusion amongst farmers about the details of their PES contracts suggests that capacity-building could have been stronger in some cases. In Uganda, for instance, the fact that payments were going to end in year seven of a 100-year contract was not fully understood by many of the participants.

Social Capital

Strong social capital can help to reduce the transaction costs of PES schemes as well as providing other social benefits (see discussion on financial capital above). Most of the PES projects in this volume appeared to work closely with existing local institutions rather than develop new institutions as part of project activities. Where collective contracts were used, existing community institutions were typically the contracting party, and the PES schemes often involved activities to strengthen local resource management and social coordination capacity (for example, Brazil, Philippines, Mexico). Even where schemes involved individual contracts, they often worked with coordinating bodies such as farmers groups (for example, Indonesia, Mozambique, Uganda). This strategy of working with and strengthening local coordination capacity perhaps underpins the finding that PES schemes generally contributed to a sense of social cohesion amongst participants.

The expansion of farmers' networks beyond the community – their bridging social capital – was also facilitated by several of the PES schemes. The Brazilian case study, for instance, reports the integration of participants into other state projects and development initiatives. Other cases also document greater community engagement with local government, stronger local participation in planning processes and greater negotiating capacity or political voice amongst participating groups (for example, Indonesia, Mozambique, Philippines). Choosing to work with and strengthen the capacity of, existing community-level institutions may be the most likely reason for this outcome.

Physical Capital

Where PES income went to community-level institutions, for instance, in Mexico and the Philippines, there was evidence of investment in community infrastructure, such as improvements in water supply, roads, clinics and schools. In the Filipino case, where the infrastructure was built a decade ago, there was evidence that the community had continued to maintain the infrastructure built using project income, reflecting a high degree of ownership on the part of the community.

Natural Capital

Most of the studies noted improvements to the status of natural resources. However, the scale of impact was typically small and its relationship to actions taken under the PES initiatives was not always clear or direct. This attribution problem arises because very little monitoring of impacts was undertaken, and when it was, it typically focused on proxies – for example, land use – for the environmental services being targeted, which might be one of a range of factors influencing a particular environmental service. For instance, the linkage between forest cover and water flow rates in streams, discussed in the Indonesian case, is complex and can be influenced by a range of factors, such as climate and water absorption by vegetation. The scale at which monitoring took place often made it difficult to assess whether leakage was occurring, a point noted as particularly relevant in the Brazilian case.

Direct and indirect impacts on informal use of natural resources are a relevant area for investigation in the planning and implementation of PES arrangements. The Ugandan study examined the issue of changing access conditions to fallow lands used by poor households for collecting fuelwood, grazing and that were sometimes borrowed for cultivation. The study found that individual smallholder contracts for afforestation reduced this informal access by poor households to fallow land, while in the Philippines, fire protection activities affected non-participants through the loss of new pasture, formerly created by uncontrolled fires.

IMPLICATIONS FOR THE DESIGN OF REDD ACTIVITIES

In order to focus attention upon the implications for livelihoods of the design of REDD activities, the following discussion is organized around the critical factors – summarized in the introduction – that influence the risks and opportunities for livelihoods arising from PES schemes.

The Nature and Location of the Environmental Service

By definition, REDD activities focus on forest areas, which are normally located in remote areas and have a high percentage of poor households (for example, Sunderlin et al. 2005). PES schemes have been shown to be able to generate beneficial impacts on livelihoods, as discussed above. Therefore, REDD activities that use PES mechanisms have the potential to contribute to the livelihoods of the poor. The extent to which this will eventuate depends on a number of factors. Here we consider the issue of targeting; other factors are discussed below.

Wunder (2009) recommends targeting areas with the highest deforestation threat, such as those near markets and roads. His argument is that only by targeting PES in such areas will a real contribution to climate change mitigation be made, since their conservation will have an additional impact on the baseline scenario. The analysis of the GEF's PES schemes, however, recommends considering social targeting in the design of PES schemes to maximize livelihood benefits. The targeting of areas of high deforestation will result in reduced emissions only in the absence of leakage. If leakage is likely to occur, other areas at lower risk of deforestation may need to be targeted as well. Mechanisms to reduce leakage have been debated in the international REDD context (Mollicone et al. 2007) but need more detailed research in relation to national-level implementation (for example, Irawan and Tacconi 2009). The targeting of high threat areas, and of more remote areas, has implications for livelihoods. Areas subject to high deforestation threats can be expected to have a lower percentage of poor people than more remote areas (that are less threatened, have more limited access to markets and, therefore, present fewer economic opportunities). National-level REDD planning will need to take into account, therefore, environmental parameters (such as the threat of deforestation and carbon densities), social parameters (such as location and degree of poverty), as well as economic parameters (such as opportunity costs, discussed below).

Resource Rights

Clearly defined community or individual rights to resources is perhaps the most complex issue facing the implementation of PES schemes within REDD activities. This leads Wunder (2009) to define the necessary condition for PES to be the identification of 'land stewards with reasonably good control over clearly delimited lands' (p. 221). There are two important aspects to the issue of resource rights in relation to the implementation of REDD by PES activities and its implications for livelihoods: (1)

the extent of forest land owned by the state, and (2) conflicting claims over ownership and use rights on forested land (often occurring on state lands). These two aspects of ownership and use rights are sometime conflated in PES discussions. Clear resource rights are a necessary condition for PES schemes, but the state has these rights over extensive forest areas that may become eligible for REDD and there are conflicting claims to many of these areas. So there is a focus in the literature on loosening state ownership and resolving conflicting claims. Intermediate solutions – such as have occurred in some of the cases discussed in this volume – where contracts can be based on partial rights – are overlooked. It is therefore important to address this issue clearly.

In many developing countries, the state owns the largest share of forest land (Table 1.1). This may be seen as limiting the applicability of PES in the implementation of REDD, given that PES contracts are typically required to be made with the land owner, which in these cases is the state.

The cases in this volume illustrate that PES schemes could, however, be relevant in three contexts: (1) if land tenure reforms devolve property rights (for example, in the case of indigenous lands), (2) if communities are given partial (for example, use or management) rights to forests and/or forest carbon or (3) if individuals and/or communities are paid for their input to the conservation of forests while the forests remain under formal state ownership.

Tenure reform has been advocated in order to make the implementation of REDD effective, equitable and efficient (Sunderlin et al. 2009). It is important to stress that such tenure reform could include either changes in the ownership of land or changes in the use and/or management rights over forests and their products. Transferring land rights from the state to communities (that is, to common property ownership),[1] would be the best option from the perspective of rural communities, because it enables more choices over the use of forest land (including conversion to agriculture and other land uses, unless enforceable restrictions on the use of the forest are attached to the rights). However, this option is politically charged, costly, and can only be implemented over a relatively long period of time. It is therefore the option least likely, at least in the short term, to be implemented by states. An alternative option is the transfer of rights to the use of forests and the carbon contained therein (for example, Streck 2009). Both options would allow the implementation of PES schemes for REDD.

In relation to who holds the rights discussed above, the design of PES schemes has mostly emphasized the role of individual landholders. Case studies in this volume (for example, Mexico and Brazil) demonstrate the viability of PES schemes focused on common property resources.[2] A PES scheme that involves communities instead of individual landholders in the implementation of a PES scheme for REDD would reduce transaction costs (further discussed later) and, by building on local community institutions (and if necessary supporting new ones), could strengthen social capital.

A third option considered in this study is paying communities, or individual community members, for their involvement in the conservation of state forests. In the Filipino case study, communities were involved in a PES-like scheme without having rights over forest carbon: the local government received payments for avoided fires and those payments benefited local communities through improved services and infrastructure (as chosen by the communities). Although there were several problems with that scheme, it shows that REDD activities could include a PES mechanism even without changing rights over forest carbon. A mechanism that does not provide funding directly to individuals and/or communities is not considered a PES scheme according to the definition of PES proposed by Wunder (2005). It would, rather, be classified as an intergovernmental transfer mechanism (for example, Kumar and Managi 2009), but there are good reasons for revising the definition of PES to include this type of payment. This extension of PES would contribute to addressing some of the restrictions inherent in the definition presented by Wunder (2005) and noted in Chapter 1.

Land tenure conflicts are widespread in tropical forest countries (Sunderlin et al. 2009) and particularly in forest frontiers where deforestation activities take place (Wunder 2009). Undoubtedly, conflicts over land rights and access to resources need to be resolved in order to implement PES schemes aimed at paying individuals or communities for providing environmental services. The need to address conflict does not imply, however, that land tenure and/or access rights reform needs to be implemented across the whole country before a PES for REDD scheme is implemented. The problem could instead be addressed on a case-by-case basis as PES schemes are developed. In the case of oil palm development in Indonesia, for instance, smallholder oil palm development has been widespread since at least the 1990s (Potter and Lee 1998), taking place in a piecemeal way despite uncertainties over land tenure and, at times, conflicts over access to land. This demonstrates that, when development opportunities present themselves, those conflicts can be resolved, on a case-by-case approach.[3] The presence of tenure conflicts is not necessarily an insurmountable obstacle to REDD activities implementing PES schemes, but it is certainly a factor that increases the social and sustainability risks and the transaction costs.

Payments

In relation to the level of payments to be provided by a PES scheme, several case studies (Mexico, Uganda, Philippines) demonstrate that there is often a failure to assess the opportunity costs faced by participants. Failure to at least match these opportunity costs results in negative impacts on livelihoods and non-compliance with contracts. It is therefore imperative that REDD activities measure these opportunity costs and seek to provide benefits at least equal to opportunity costs and the transaction costs faced by participants (for example, time input to meetings).

One way of setting prices that address opportunity costs is through the use of auctions (Jack et al. 2009), in which resource rights holders bid to offer their resources for inclusion in a PES scheme at a certain price. Auctions have also been recommended for REDD-related PES schemes (Wunder 2009). The application of auctions may be constrained, however, in PES for REDD by the resource rights conditions discussed above. Auctions are most appropriate where the suppliers of services are individual landholders. As already noted, most forests that may become eligible for REDD activities are state owned and, as such, community participation may be limited to involvement in conservation activities on state land.

Two significant issues need to be addressed in relation to payments which also affect the appropriateness of using auctions, even in the context of private or communal forests. First, details of how REDD will work are yet to be agreed by the Conference of the Parties to the UNFCCC, but REDD activities will be expected to lead to the long-term conservation of forests, over the next 50 to 100 years. Communities and households entering into PES contracts would therefore lock up the forest and land for many years in the face of great uncertainty about future returns from the land and forest. To avoid imposing all of this livelihood risk on local resource rights holders, PES contracts should allow for a regular review of the payments provided in exchange for the environmental service to ensure that opportunity costs were covered over the life of the contract. The development of appropriate review mechanisms would need to consider whether auctions could be used at the established review time. If auctions could not be accommodated, there would be a need to assess whether they lead to sufficient efficiency gains to justify their use at the beginning of the PES scheme.

Second, how the schedule of payments matches the length of the contract is another issue that PES schemes for REDD activities need to address. The financial benefits have often been completely disbursed during the first decade of a PES scheme (for example, Mozambique), but participants are sometimes not aware of the terms of the contract and think that the end of the payments signifies the end of the contract, which will then be renewed. This misunderstanding could generate significant problems for the viability of REDD activities, which would normally be expected to last for several decades. The payment stream should be designed to last for the duration of the contract to ensure conditionality of payments and to benefit livelihoods.[4] However, aligning the payment schedule with the contract length has implications for the price of carbon, which is much lower for short-term credits.

REDD activities are expected to conserve forests over long periods of time, thus affecting future generations, who can be expected to seek benefits from the form of land use agreed in the contract. Cash payments would not deprive future generations of their rights to benefit from resources if the payments are scheduled to take place over the life of the activity and if the contracts include options for the revision of the amount paid in order to account for changes in opportunity costs that can be expected to occur over long periods of time. A similar argument applies to payments through the provision of services and infrastructure. Infrastructure cannot necessarily be expected to benefit future generations given that it deteriorates over time, therefore appropriate maintenance would need to take place.

There is also a need to further explore issues around benefit distribution, for instance, individual versus collective payment and management of benefits. Several case studies (Mexico, Brazil and Philippines) show that collective payments are a viable option, particularly where there are existing and functioning institutions for collective action. However, where such institutions do not exist, individual benefits or a combination of approaches may need to be considered. Whether payments are made to households or communities and whether they are paid in cash, through provision of services, or a combination of the two, will also need to be considered by PES schemes for REDD.

The preferences of the participants should be considered in deciding whether the payments are made to individual households and/or to communities. For instance, when households commit their own resources (for example, land) to the provision of the environmental service, payments directed to individuals are clearly appropriate. When communities commit shared resources, consideration of the provision of community-level benefits may be appropriate. These benefits could be in cash paid to households who participate in the REDD activities (for example, involvement in monitoring forest conservation) and to households who do not participate directly but who may be affected by the conservation activities (for example, through loss of access to forest products). Benefits might also be delivered by providing infrastructure and services. Whilst the latter could be expected to be preferred in situations involving community conservation activities, it may also be chosen by individual participants who find it difficult to access certain services, for example, training, as demonstrated by the case of the GEF's PES schemes. Cash payments, or the provision of services, could also be appropriate benefits for involvement in conservation activities by individuals and communities residing in, or managing, state forests.

The conditionality of payments is a fundamental characteristic of PES schemes (Wunder 2009) and affects the permanence of the carbon stock. There has been, however, a significant difference between the theory and the practice of PES implementation: PES schemes have so far given limited attention to monitoring and enforcement of agreements, as demonstrated by the case studies in this book. Given that the permanence of the carbon stock is fundamental to REDD activities, the establishment of baselines and the monitoring of the services (that is, carbon) will be an integral part of the PES schemes for REDD. However, whether PES schemes for REDD will be able to ensure permanence is an open question, which depends on human as well as natural events, such as wildfires. In the event of non-permanence, it is unlikely that payments provided to communities could be recovered. In some cases, it may be inappropriate to seek to recover those funds because the events may have been beyond the control of the individuals or communities involved in the scheme. There are obviously risks that need to be addressed and ways of insuring against them need to be developed, while at the same time ensuring that they do not generate moral hazard (such as forests being cleared for agricultural development by local communities using intentional fires which are made to look like accidental ones).

Finally, the literature on PES has focused on the need to monitor the flow of the environmental service. However, the flow of payments also need to be monitored because corruption can be expected to affect REDD activities including the related PES schemes, as also demonstrated by the case study in Indonesia. If corruption were to affect the flow of payments, the livelihoods of the participants would be negatively impacted and the continuation of the PES scheme would be at risk.

Participation in PES Schemes for REDD

Participation of (poor) rural households and communities in PES schemes for REDD is possible, as demonstrated by the case studies. However, several factors will continue to influence the participation of poor households. The issue of ownership and other forms of resource rights is the most fundamental factor, and has been discussed above. Other factors, including the skills, education, power and negotiating participation of potential participants, transactions costs and the availability of credit to facilitate participation are considered below.

The skills, education, power and negotiating capacity of rural people have been found to affect their ability to participate in, and the benefits they receive from, PES schemes. The role of intermediary organizations in PES schemes has been underestimated (Vatn 2010), but they have been playing an important role in addressing these constraints, as highlighted in several case studies presented in this book. Rolling out national REDD activities would, however, require a considerable number of intermediary organizations to become involved in the related PES schemes. Whether a sufficient number of intermediary organizations have the required expertise to carry out the necessary activities is an issue that will need to be considered in the development of national REDD programmes. Intermediary organizations without this expertise can have significant negative impacts on project implementation, as demonstrated in the Mexican case study, so it may be necessary to conduct capacity-building activities for organizations facilitating PES schemes for REDD.

Transaction costs can be a significant constraint upon the participation of rural people in PES schemes as noted, for example, in the case study on Colombia and Nicaragua. A REDD activity seeking to engage a large number of smallholdings would face higher transaction costs compared with a REDD activity that dealt with few large landholdings. Private proponents of REDD activities are therefore more likely to focus on the latter type of activity, at a similar level of opportunity costs. Transaction costs of dealing with smallholdings can be reduced by setting up contracts that involve groups of smallholders rather than individuals. Group contracts with smallholders work best when payments are received by individual households, but this means the transaction costs are not reduced as much, as discussed in the Colombian and Nicaraguan case study. Another way to reduce transaction costs is to develop REDD schemes that combine carbon sequestration on individual land holdings with sequestration in community forests, as discussed by the case study on Mozambique, thereby increasing the total volume of carbon sequestered by that REDD activity.

Finally, REDD activities which seek the involvement of rural people in the enhancement of carbon stocks (which is part of REDD-plus and may involve planting trees) will need to consider the finance and credit availability faced by those people to cover their up-front costs of participation. These up-front financial resources would be difficult to recover in the case of non-compliance with contracts, and appropriate insurance mechanisms would need to be considered by those providing the funding.

In conclusion, the evidence from the case studies in this volume indicates that PES schemes can have positive livelihood impacts if they are carefully designed and implemented. A number of lessons can be learned from the experience of PES schemes, particularly with respect to maximizing positive environmental, social and economic impacts. Lessons relate to the recognition that trade-offs often need to be made in achieving both environmental and social objectives, and the lack of secure resource rights can constrain access to PES schemes for poor households. On the other hand, the cases have illustrated options for PES to proceed in the absence of full ownership rights to resources. Other factors affecting the ability of the poor to participate include their skills, education, power and negotiating capacity, all of which will need to be considered in the design of REDD projects. The need to consider opportunity costs when setting prices was also identified, as failure to at least match opportunity costs results in negative livelihood impacts and long-term failure of PES schemes. Payment schedules also need to be carefully designed to benefit livelihoods and need to be tied to monitoring processes to ensure conditionality of payments. Whether these payments are made to individuals or communities is another key choice with implications for benefit flow and transactions costs.

NOTES

1. We do not consider transfer of land rights to individuals (private property) because this is not proposed in the existing literature on REDD and there would be considerable problems with its implementation.
2. This does not imply that the design of the PES scheme considered in the Mexican case study is perfect.
3. This is not meant to imply that oil palm developments, particularly large-scale plantations, have not been the cause of conflicts (for example, Harwell 2000).
4. See Tacconi and Bennett (1995) for an example of a payment stream designed over the life of the conservation activity.

REFERENCES

Corbera, E., K. Brown, and W.N. Adger (2007a), The equity and legitimacy of markets for ecosystem services, *Development and Change* **38** (4), 587–613.
Dev, O.P. and J. Adhikari (2007), Community forestry in the Nepal hills: practices and livelihood impacts, in O. Springate-Baginski and P. Blaikie (eds), *Forests, People and Power. The Political Ecology of Reform in South Asia*, London: Earthscan, pp. 142–76.
Harwell, E. (2000), Remote sensibilities: discourses of technology and the making of Indonesia's natural disaster, 1997–98, *Development and Change* **31** (1), 307–40.

Irawan, S. and L. Tacconi (2009), Reducing emissions from deforestation and forest degradation (REDD) and decentralized forest management, *International Forestry Review* **11** (4), 87–98.

Jack, B.K., B. Leimona and P.J. Ferraro (2009), A revealed preference approach to estimating supply curves for ecosystem services: use of auctions to set payments for soil erosion control in Indonesia, *Conservation Biology* **23** (2), 359–67.

Kumar, S. and S. Managi (2009), Compensation for environmental services and intergovernmental fiscal transfer: the case of India, *Ecological Economics* **68** (12), 3052–59.

Landell-Mills, N., and I. Porras (2002), *Silver Bullet or Fools' Gold? A Global Review of Markets for Forest Environmental Services and Their Impact on the Poor*. London: International Institute for Environment and Development.

Pagiola, S., A. Arcenas and G. Platais (2005), Can payments for environmental services help reduce poverty? an exploration of the issues and the evidence to date from Latin America. *World Development* **33** (2), 237–253.

Pagiola, S., A.R. Rios, and A. Arcenas (2008), Can the poor participate in payments for environmental services? Lesson from the Silvopastoral Project in Nicaragua. *Environment and Development Economics* **13** (3), 299–325.

Potter, L. and J. Lee (1998), Tree planting in Indonesia: trends, impacts and directions, Occasional Paper No. 18, Bogor: Center for International Forestry Research.

Streck, C. (2009), Rights and REDD+: legal and regulatory considerations, in A. Angelsen (ed.), *Realising REDD+: National Strategy and Policy Options*, Bogor: Center for International Forestry Research, pp. 151–62.

Sunderlin, W.D., A. Angelsen, B. Belcher, P. Burgers, R. Nasi, L. Santoso and S. Wunder (2005), 'Livelihoods, forests, and conservation in developing countries: an overview, *World Development* **33** (9), 1383–1402.

Sunderlin, W.D., A. Larson and P. Cronkleton (2009), Forest tenure rights and REDD+: from inertia to policy solutions, in A. Angelsen (ed.), *Realising REDD+: National Strategy and Policy Options*, Bogor: Center for International Forestry Research, pp. 139–50.

Tacconi, L. and J. Bennett (1995), 'Biodiversity conservation: the process of economic assessment and establishment of a protected area in Vanuatu, *Development and Change* **26** (1), 89–110.

Vatn, A. (2010), An institutional analysis of payments for environmental services, *Ecological Economics* **69** (6), 1245–52.

Wunder, S. (2009), Can payments for environmental services reduce deforestation and forest degradation?, in A. Angelsen (ed.), *Realising REDD+: National Strategy and Policy Options*, Bogor: Center for International Forestry Research, pp. 213–23.

Wunder, S. (2008), Payments for environmental services and the poor: concepts and preliminary evidence. *Environment and Development Economics* **13** (3), 279–297.

Index